CURD & CRUST

ARTISAN CHEESE AND
BREAD MAKING

Tamara Newing

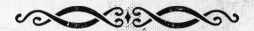

This book is dedicated to the memory of my precious son, Reagan Milstein, who was taken from us tragically at age 14.

Cheese was Reagan's favorite food and making cheese in all its delicious varieties helped me navigate the grief following his accident.

My royalties from this book will be donated to Reagan's charity which raises money for underprivileged children, to help improve their lives and opportunities through sport.

www.rmf.world

CONTENTS

INTRODUCTION	9
CHEESE MAKING BASICS	11
A note on how to use these recipes….	11
CHEESE MAKING TECHNIQUES – STEP BY STEP	13
EQUIPMENT FOR THE NOVICE CHEESEMAKER	31
INGREDIENTS FOR SUCCESSFUL CHEESEMAKING	33
CHEESE	39
Cow's milk cheese	41
Goat's milk cheese	65
Buffalo and sheep's and mixed milk cheese	86
BREADS	103
COOKING WITH CHEESE	125
GLOSSARY	157
INDEX	159

INTRODUCTION

For centuries, cheese has been made in communities around the world.

In its earliest known form, cheese was created almost by accident. It is believed that nomads happened upon the magic of coagulation when they transported fresh milk in bags formed from the lining of an animal's stomach. As the milk was transported in the desert heat, it began to coagulate and when the 'bag' was opened, the milk had magically transformed from liquid into a soft curd. This was the first known use of rennet, an enzyme which exists naturally in a mammal's stomach.

Over the centuries, vegetable rennet has also been discovered in the sap of fig trees and even in these modern times, it isn't unusual to see an elderly woman in the hills of Greece, clad in black from top to toe, stirring her cauldron of milk over an open fire and breaking a branch off a nearby fig tree to stir and coagulate the milk.

For modern cheesemakers, we can now choose animal rennet, vegetable and microbial rennet.

You should also feel very comfortable using the most rudimentary items in your kitchens. Early cheesemakers didn't have thermomixers, bain maries or stainless steel draining benches and while these are lovely to use, they are certainly not necessary. I encourage my students to use what they have in their kitchens and the 'equipment' section of this book will give you some guidelines.

Importantly, you should understand that cheese making is a journey of a thousand steps. I hope that this book will give you the basics and the confidence to take that first step and that once you are on the cheese-making road, the thrill of producing cheese and sharing the delicious results sees you taking further steps down that road.

I've been making cheese now almost every day for six years and I still find the transformation of milk to curd exciting and incredible. Yes, there are frustrating days too, when things don't go to plan, but I have honestly never wanted to stray from this path. Each batch helps me perfect my skills and I never stop learning.

With a little confidence, you will find that you will be able to stamp your own individuality on your cheese, adding herbs and spices, making different shapes, aging them, washing them and drying them to create your own perfect cheese.

Together, we can take the first step down that long, happy, sometimes frustrating and absolutely delicious cheese-making road.

Bon Appetit!

CHEESE MAKING BASICS

People have been making cheese for literally thousands of years in rudimentary conditions much less convenient than those we take for granted today. We have access to kitchen utensils, running water and quality, clean milk. With a few basic guidelines, everyone can make successful, healthy and delicious cheese!

Good hygiene is the most important factor in learning to make cheese. Aside from washing and sanitising your hands, all utensils, draining moulds, draining trays, etc., must be sanitised before use. This kills bacteria that could have a negative impact on your cheese, its lifespan and its flavor. The most convenient sanitiser for home cheesemakers is the kind used to sanitise baby bottles. Liquid or tablet form is fine. Just follow instructions for dilution on the packet. If you have items to sanitise that are too large to submerge, put some sanitiser in a clean spray bottle and spray the items, then wipe with kitchen paper.

Remember, cheese is a living food on which we aim to grow 'good bacteria' such as the white mould on a Brie or the blue mould in a Stilton or Roquefort-style cheese. Once the cheese is transferred to its humid environment to grow these healthy moulds, harmful bacteria can also thrive. Sanitising your utensils and keeping your hands sanitised will ensure you don't introduce them into the cheese-making area.

A NOTE ON HOW TO USE THESE RECIPES…

As a teacher first and cheesemaker second, my aim is to provide you with easy, approachable recipes that demystify the process of cheese making and enable you to successfully create wonderful cheese at home.

As such, I won't delve too deeply into the science of cheese making and give you pages of notes. If you are the type who always wants to know how and why things do what they do, there are many wonderful books available on the science of cheese making and the Internet is also a fabulous resource.

Each recipe is complete and gives you all the information you require, but some will also direct you to a particular technical step that may explain a necessary process in more detail.

This book is designed to give you enough information for successful cheese making but not so much that you feel overwhelmed before you start!

Happy cheese making!

CHEESE MAKING TECHNIQUES – STEP BY STEP

Most cheese making recipes follow the same principles, though some, such as Ricotta and Paneer, are heated, acidified and strained quickly.

If you are making a cheese with rennet, you will follow some or all of these steps:
1. Warming the milk
2. Adding the starter cultures
3. Adding the calcium chloride, if using, and then the rennet
4. Setting the curd and checking for 'clean break'
5. Cutting the curd
6. 'Healing' the curd
7. 'Cooking' (heating) the curd (for hard cheeses only)
8. 'Wheying off' and draining the curds
9. Moulding the curd
10. Milling, salting, pressing the curds (for hard cheeses only)
11. Turning the cheese
12. Salting or brining the cheese
13. Ripening/aging the cheese

Each recipe contains all the necessary steps for each particular cheese but in this section, we explain in more detail the steps that require a little more detail.

➦ 1. Warming the Milk

Your recipe will tell you what temperature to heat the milk to. You can heat it in a saucepan but milk-setting temperatures are low and it is easy to over-heat.

I prefer to put the milk in a sanitised container and place it in a sink full of very hot or boiling water. Stir the milk gently and you will find that it will reach the required temperature quickly. Remove the container from the water and proceed with the recipe.

If you do overheat the milk, sit it in a sink full of cold water and allow it to slowly come back to the required temperature. Do not be tempted to set the milk at a temperature more than 1°C hotter than the recipe stipulates as this will have an impact on the resulting cheese.

➤ 2. Adding the starter cultures

Most starter cultures these days are DVS (Direct Vat Set) which means you add them directly to the warmed milk and stir to dissolve. Once dissolved in the milk, they will begin to multiply and acidify the milk. I like to leave my milk for a minimum of 20 minutes before I proceed with the recipe, unless this is a 'lactic set' cheese in which case I will leave for an hour because the milk is cooler so the cultures take longer to acidify. If your recipe calls for secondary cultures like white mould, blue mould, geotrichum or brevi linens, you will add these now.

If you prefer to use your cultures to make a 'mother culture', you will add the cultures to 1 litre (35 fl oz/4 cups) of warmed milk, stir well and then leave to set in a warm place overnight until the mixture looks and smells like yoghurt. The following day, you will add half this mixture to your full amount of milk in your recipe and proceed as per the recipe. The remaining 2 cups of 'mother culture' can be frozen for another cheese make or used in another recipe within two days. Keep any unused 'mother culture' in the fridge until using.

➼ 3. Adding Calcium Chloride and Rennet

When you are ready to add the calcium chloride and rennet, you will need a sanitised syringe so that you can add exactly the amount you need. Add the calcium chloride first and mix gently but thoroughly. Then add the rennet and mix thoroughly for no more than 2 minutes. After this time, the rennet will begin to set and the milk must be left alone and not moved or touched during this time, approximately one hour.

➤➤ 4. Setting the Curd and 'Clean Break'

Once you have allowed the milk to set, it should look a little like a large block of silken tofu. If you give the container or vat a gentle shake, you should see the mixture wobbles a little but is clearly no longer in liquid form. When fully set, you should notice a clear liquid forming on the surface. It may be a drop of two or a thin film. This is whey and is an indication that the milk is fully set and you are ready to proceed.

To check, take a clean and sanitised (preferably plastic) knife and make a shallow cut in the curd, then turn the knife blade flat and lift the curd where you made the knife cut, to see if the cut lifts on the blade of the knife in a firm slice. If the curd cuts and lifts cleanly, it is ready. If you notice the curd disintegrates on the knife or seems to be more like custard than tofu, leave for a further 10 minutes and check for 'clean break' again.

➽ 5. Cutting the Curd

Once you have ascertained that the curd is fully set, it is time to cut the curd. The size of the cutting will depend on the type of cheese you wish to make. For example, a brie or soft-ripened style cheese recipe will instruct you to cut the curd into large squares about 1 cm (½ inch) and for a cheddar, Tomme or other semi-hard or hard cheese, the cutting will continue until the pieces are much smaller—pea size or even rice size pieces of curd.

The smaller the pieces, the harder the cheese. Some recipes will instruct you to cut the curd into larger pieces and then when reheating the whey, continue cutting into the smaller pieces.

➦ 6. 'Healing' the Curd

Once the curd has been cut to the desired size, whey will immediately be visible as it accumulates in the vat. To 'heal' the curd, I stir it gently after cutting and then allow it to rest for between 5–10 minutes. This allows whey to continue to be expelled from the curd but also allows the surface of each piece of curd to firm up. When you are ready to move to the next step, the curd will likely have sunk to the bottom of your vat, completely covered by the accumulating whey. Give the curd a brief stir to loosen it and you are ready to move on.

➦ 7. 'Cooking' and/or further cutting the curd (for hard cheese only)

If you are making a hard cheese, you will now be instructed to add some heat to the vat to increase the temperature of the whey by about 4–5°C (10°F). If you are using a plastic vat, immerse it gently in a sink of boiling water and stir the floating curds until a thermometer registers the required temperature. Remove the vat from the water. If you are using an electric vat, gently increase the temperature and stir the curd until it is a little LESS than the temperature you require. Turn off the heat and allow the curd to reach the desired temperature with the residual heat. If the recipe stipulates, you will now also continue to cut the curd to smaller, rice or pea size pieces.

➼ 8. 'Wheying off' and Draining the Curd

Commercial cheese producers have vats with a tap where the whey can be drained, leaving the curd in the vat for hooping and shaping. For home cheesemakers, I find it easier to remove some of the whey before I shape the curds into cheese. Removing most of the whey makes it less messy but also enables you to see how much curd you have to work with.

Place a sanitised plastic colander or strainer onto the surface of the curds and wait a moment or two for the colander to fill with whey and to sink a little. Using a sanitised jug or small bowl, scoop out the weigh and discard (or save for another use—see 'Uses for leftover whey' at the end of this section).

Continue scooping and removing the whey until the colander no longer fills with liquid. You should now be easily able to see the curd and it should be no longer covered with whey. Carefully lift the colander and see whether it has left an indentation on the curd. If so, your curd is coagulating and is ready to be hooped immediately. If there is no indentation, stir the curd gently and wait a further minute or two before proceeding.

➻ 9. Moulding the curd

You should now have sanitised your chosen cheese moulds and will be transferring your curd to the moulds. If you are making a soft cheese such as a Brie style or lactic cheese, you will gently scoop the curd and transfer to the mould, tapping down gently to fill any gaps and allow the cheeses to drain gently.

If you are making a hard cheese, you will transfer the smaller, firmer curds to their moulds and press down firmly, often by placing one cheese mould as a weight on another and then after a few minutes, reversing them from top to bottom so that each cheese has a chance to be weighed down.

➼ 10. Milling, Salting, Pressing the curds (for hard cheeses only)
If you are making a hard cheese, you will now be instructed to drain the curds completely by transferring them from the vat to a large colander. I like to line the colander with sanitised cheesecloth to prevent the loss of any tiny pieces of whey. Once the curds are completely drained, your recipe may ask you to use your hands to further break up the clumps of curd, to add salt and mix, or to gather and press the curds firmly to dispel more whey.

➳ 11. Turning the Cheese

Once your curds have been placed in their moulds, they will be there for a period of between a day and several days, depending on the recipe. Within the first hour, you will turn the cheese by gently tipping it out of the mould onto your sanitised hands, turning it and returning it to the mould upside down. Sometimes the cheese will be difficult to remove the first time but gently ease or tap it out. Don't be tempted to force it.

The reason for turning the cheese is to give it an even shape, top and bottom but more importantly to allow even whey loss so that the cheese has the same consistency throughout. Your recipe will tell you how often to turn your cheese. Typically, you would turn it within the first hour, after 2 hours, after another couple of hours and then again the following day.

➼ 12. Brining or Salting your cheese

Salt acts as a preservative for your newly made cheese and also adds flavor. Cheese is typically salted or brined on the day after it is made. All cheese can be brined or salted but it is generally accepted that blue cheese prefers the aggressive nature of dry salting while softer cheeses do better when brined.

To Brine Your Cheese: Make a saturated brine by mixing a cup of cooking salt (non iodised) with 1 litre (35 fl oz/4 cups) hot water. Mix well to help the salt dissolve, then cool overnight, then chill. This is best done on the day you make your cheese. The day after your cheese is made, remove your brine from the fridge and gently submerge your cheese in the brine for approximately 2 hours for each 300 g (10½ oz) weight. For example, if your cheese weighs 150 g (5½ oz), you will only soak it in the salt brine for 1 hour. If it weighs 600 g (1 lb 5 oz), you will soak it for 4 hours. (Add more time if you prefer your cheese salty, less if you don't.)

If the depth of the salt brine is not sufficient for the cheese to be submerged, turn the cheese over halfway through the required time so that the cheese is evenly brined on both sides. After the required time, remove the cheese from the brine and allow to drain overnight on a sanitised wire rack or cheese mat, making sure the cheese is not sitting in its own liquid. The brine can be saved for future use, stored in the fridge. Before using again, strain to remove any sediment and add ½ a cup of salt and mix thoroughly. Do not worry if all the salt doesn't dissolves. This simply means your brine is saturated with salt.

To Dry Salt Your Cheese: As a general rule, cheese is salted at a ratio of 2%. You will need a set of digital scales to accurately weigh the salt. First, remove your cheese from its mould and weigh your cheese. Multiply its weight by 2%. For example, if your cheese weighs 200 g (7 oz), you will require 4 g (¼ oz) of salt. Once you have accurately weighed the salt, sprinkle it all over the cheese and gently rub your sanitised hands over the cheese, making sure the salt is evenly distributed and taking care not to lose too much of the salt. Place the salted cheese on a wire rack or draining mat until the following day (you can return the cheese to its mould if you wish).

➺ 13. Ripening and Aging Your Cheese

Ripening is the stage that takes place as the mould grows. Once the cheese is salted and drained for a further day, it is time to commence the ripening process. If your cheese is going to develop either a blue or white mould, it is going to need humidity and warmth. Ideally, the cheese should live at approximately 10–12°C (50–54°F). A wine fridge or cellar works well but if you don't have access to either of these, a domestic fridge will suffice. To manage the humidity, I suggest putting the cheese in a plastic container that has been lined with a fresh piece of paper towel, then covered with a plastic draining rack (many food containers now come with a plastic rack in the bottom, or cut a piece of cheese draining matting to fit the base of the container). Arrange the cheese on the sanitised mat and replace the lid. Leave the box of cheese in your chosen location for 24 hours.

The next day, remove the lid and check that there is 'fog' or condensation under the lid. The lid should not be too dry (the lid wasn't on properly) or too wet (the lid needs to be left ajar to allow a little air circulation).

If you are super attentive to detail, you can buy a thermometer/agrometer and check the levels of both temperatures and humidity in the box.

Once you are comfortable that the temperature and humidity are correct, the cheese should be turned every 2 or 3 days to ensure even mould growth, and the paper changed when it is too wet (damp is ok). At 10–12°C (50–54°F), the mould should be fully grown after two weeks. In a domestic fridge, 3–4 weeks.

If you are making a washed rind cheese, you will need to wash/brush the cheese with liquid. Some recipes call for a light salt brine, beer, wine or spirits. Other recipes call for a 'bacterial brine' which is made by adding a secondary culture called Brevi Linens with cooled boiled water. This helps to create the typical orange glow on washed rind cheese. Regardless of your chosen liquid, you will brush the cheese all over using a sanitised pastry or makeup brush until the cheese is damp but not wet. It is then returned to the aging box. Each time you turn the cheese, you will also wash it.

Once the cheese is fully ripened according to your recipe, it can be 'aged'. This allows the young cheese to fully develop its required flavors and texture. Once you have ascertained it is fully ripened and covered with its mould, gently pat the mould back to compact it then wrap the cheese in perforated cheese paper or parchment paper. If using parchment paper, place the wrapped cheese in a plastic container and chill for required time (2–3 weeks for a soft cheese and up to several months for a hard cheese).

➤ Uses for leftover whey

The whey from a lactic cheese is very acidic and I find its only use is as the liquid in a bread recipe instead of water. This creates quite a sour dough and as the words suggest, makes a great, 'cheats' super-quick sourdough type bread.

For whey from Brie, cheddar and other rennet-set cheese, the following ideas will help you use up the high protein liquid:

- Bring whey to the boil with a small amount of fresh milk (about 20% volume of the milk) and when about to simmer, you should see particles of curd rise to the surface. Turn off the heat and wait 5 minutes before scooping your 'whey ricotta' into a draining basket. The remaining whey can be used in in soups and bread recipes or cooled and used on a garden as a pesticide.
- If you grow 'acid loving' vegetables such as tomatoes, zucchini (courgette) or capsicum (bell pepper), a whey bath once a week when watering will give them incredible growth. Mix your whey with an equal quantity of water before watering the plants.
- Use the whey in a soup instead of water, it adds fabulous creaminess to the soup.
- If you go to the gym and want an extra burst of protein to help repair your aching muscles, flavor the whey with sugar, vanilla, chocolate or coffee and drink within two hours of a workout.

EQUIPMENT FOR THE NOVICE CHEESEMAKER

If you love kitchen accessories, cheesemaking will open up a whole new world of vats, plastic draining moulds and baskets, cheese press equipment, cheese harps, curd cutters and more. However, please keep in mind that cheese has been successfully made in rudimentary 'kitchens' with no running water or modern conveniences for thousands of years, and while these items are nice to have, they are not essential for successful cheese making.

Below is a list of items that a 'first time' cheesemaker will find useful. Most of these are regular household items and there should be no need to go out and purchase anything expensive.

Of course, once you succeed in making cheese and want to make more, you will no doubt find great pleasure in purchasing more specific items that make your cheese journey even more rewarding.

➤ **Cheese Vat** – This is the item that you will set your milk in. It can be as simple as a large saucepan (scrupulously clean and with no damaged non-stick surfaces), a plastic tub (10 litre volume or more) or even a plastic bucket. Whatever item you choose, remember it needs to be able to contain a minimum of 10 litres (338 fl oz) of milk with room for stirring. Also make sure this item can be sanitised either by immersing in sanitiser or be using a spray bottle of sanitiser.

➤ **Plastic knife for 'cutting the curd'** – I use a plastic 'lettuce knife' for this purpose but you could use any knife that can be sanitised. Don't use ones of dubious metal as they may leave a metallic 'taint' on the curd.

➤ **Curd Cutter/Cheese Harp** – This is used to cut the curd into even squares but for your first few attempts, the above lettuce knife will do a good job. It will take a little longer with the lettuce knife but persevere to get the right size curd for your recipe. You may remember the old-fashioned ice cube trays that sat on a plastic base? If so, the frame of this ice cube tray makes a perfect curd cutter for small batches 10 litres (338 fl oz) and under. You can often find these in the kitchen section of charity stores.

➤ **Cheese Moulds** – Eventually you may want to have a selection of matching moulds so that you can turn out cheese that looks uniform and perfect but these aren't essential. Take a trip to a discount or variety store and have a hunt around the kitchen section for plastic baskets, drainers and strainers. You will be amazed at what you find. Any food grade plastic item that has holes in the base and sides will work well. When I first started, I used plastic 'cutlery drainers' for my blue cheese. They are the right size and shape and have holes at the bottom,

though I did have to get my husband to make some holes for me around the sides.

➻ **Cheesecloth and Chux Wipes** – Chux Superwipes make fantastic disposable draining cloths. Make sure they are new, fresh from the packet and sanitised before use. For cheesecloth, purchase some from a haberdashery store then trim any rough edges to ensure no threads are loose. Sanitise before using.

INGREDIENTS FOR SUCCESSFUL CHEESEMAKING

Milk – Traditionally, cheese was made with raw milk, taken straight from the animal at a convenient 'setting' temperature of about 37°C (99°F) and much of the world's cheese is still made this way. However, in 'first world' countries like America and Australia, concern over food poisoning outbreaks has resulted in laws governing milk sold for human consumption and commercially sold milk must be pasteurised.

➻ **Pasteurisation:**
The process of pasteurisation aims to kill any pathogens that have contaminated the milk by way of the milking shed, or utensils used to store the milk prior to cheese making. During pasteurisation, the heat kills these pathogens but also kills the friendly bacteria that would have given your cheese its flavor and complexity. We then need to introduce cheese or 'starter' cultures to give back these friendly and flavorsome cheese bacteria. If you have your own milking animals or access to clean farm-fresh milk, you may like to make raw milk cheese. If you do have access to raw milk but prefer to pasteurise before using, refer to the techniques for pasteurisation in our 'Techniques' section.

➥ Homogenisation:

Most supermarket milk is also homogenised. Homogenisation has nothing to do with health and is a process that suspends the fat or cream in the milk, in tiny droplets rather than coagulating at the top of the bottle or carton. This process is quite aggressive and damages the cellular structure of the milk, making the curds softer and lowers your eventual yield. Most supermarkets these days also sell 'unhomogenised' or 'farm milk' which is pasteurised but not homogenised. The best milk for cheese making is milk that is as unprocessed as possible, so choose unhomogenised where possible. Do not attempt to make cheese from UHT (Ultra Heat Treated) milk as this milk has been so heavily and aggressively processed that you will be unlikely to get any curd set at all.

➥ Starters/Cultures

Cultures are the spices of the cheese world! There are literally thousands of known cultures for cheese making around the world and these add flavor and character to your cheese. These cultures put back the friendly bacteria which are lost during pasteurisation and have a direct impact on the style and flavor of your cheese.

Dozens of these cultures are commercially produced for cheese makers which is convenient but means we are all using similar strains. Many artisan cheese makers around the world are able to 'harvest' cultures and bacteria that exist naturally in their regions, and these can give their cheese a specific style and flavor which doesn't exist outside that region.

Much work is being done by various cheese associations to harvest and reproduce some of these indigenous strains so that we can all enjoy the diverse complexity of their flavors.

Once these cheese cultures are added to the warm milk, they acidify the milk and turn the natural sugars in the milk (lactose) into lactic acid, which has an impact on the style and flavor of the cheese.

Starters are either mesophilic (for temperatures up to about 39°C/102°F) and thermophilic (for temperatures above 40°C/104°F) so you need to know what cheese you are making in order to decide which cultures to use. For example, if you are making a cheddar, you will need a mesophilic culture to acidify the milk but also a thermophilic culture which manages the warmer temperatures of a curd that is further heated after cutting.

These cultures affect the flavor, curd set, acid production and aroma of the cheese and are available either as DVS (direct vat set i.e. added directly to the milk) or added to the milk and set overnight at a lower temperature to form a 'mother starter' which can then be used in the cheese making process, much like a sourdough culture for breadmaking (see Basics – Step 2).

In addition to these starters, there are also secondary cultures used to ripen the surface of the cheese. These produce white mould, blue mould, yeast surfaces and the golden/red sticky surface of a washed rind cheese, called Brevi Linens.

➻ Calcium Chloride – This is often used in cheesemaking where there is a risk that some of the calcium may have been lost during pasteurisation or homogenisation. It ensures the milk has the correct calcium to enable the curds to form, set and cut properly.

➻ Renne – Rennet is an enzyme that causes milk to coagulate. Found in the lining of an unweaned calf or kid, the earliest known cheese makers would add a small part of the animals stomach into the milk as it heated, created the world's first known rennet. These days, animal rennet is still is derived from the lining of an unweaned calf or kid but there are vegetarian versions as well, derived from fig bark as well as other vegetal sources. Microbial rennet is also available from cheese making suppliers if you prefer. These are available in different strengths so always check the correct ratio. All recipes here use single strength vegetal rennet.

Many recipes suggest diluting rennet with cooled boiled water prior to adding to the milk. This is to ensure that it is evenly mixed thoroughly into the milk. You can avoid this step but must be sure to gently and thoroughly stir the rennet through the milk for about 2 minutes before allowing the milk to set.

Rennet can be negatively impacted by heat and chlorine so follow these instructions to avoid any problems:

- Always rinse sanitised utensils with boiling water to ensure no traces of chlorine.
- If sanitising with boiling water, allow utensils to cool to room temperature before using. Heat deactivates rennet.
- If you are diluting rennet with cooled boiled water as per recipe, immediately add to milk once diluted. Do not allow to sit.

CHEESE

SPREADABLE CREAM CHEESE

Makes: Approximately 2 cups

If you've never had 'real' homemade cream cheese, you've been missing out! Unfortunately most of the commercial varieties contain stabilisers and gums to give a spreadable texture and long life but they also add a plastic texture. This beautiful, silky cream cheese has so much flavor and is so easy to make, you can do so whenever you want it!

INGREDIENTS

1 litre (35 fl oz) full cream cow's milk
300 ml (10½ fl oz) pure cream
4 tablespoons yoghurt (live culture)
1 tiny speck of mesophilic culture (just as much as will fit on the tip of a knife)
salt, to taste

METHOD

- Heat the milk, cream and yoghurt and mesophilic culture together to reach 33°C (91°F) and stir gently. Pour into a yoghurt maker or insulated cheese 'vat' (see Basics – Equipment) and allow to set, undisturbed for 24 hours.
- Once the mixture has set and is the texture of pot-set yoghurt, spoon gently into a cheesecloth-lined colander. Allow to drain on a clean draining board or sink for 12 hours, covering the exposed surface of the curd with the overhanging cheesecloth.
- Check the texture of the curd. Once it is as thick as you like, transfer to a non-reactive bowl and add salt to taste. I add 1 teaspoon for this recipe but add more or less as you wish. If you would like the cream cheese to be thicker, spoon curds back into cheesecloth-lined colander and check again in 4 hours.
- Once the curd is as thick as you desire, transfer to a glass or plastic container with a lid and store in the refrigerator.
- Use within 2 weeks.

Note: Add chopped fresh herbs, chilli, lemon zest of other flavorings as desired.

WHOLE MILK RICOTTA

Makes: Approximately 2 cups

Fresh ricotta is one of the most rewarding cheeses to make because it can literally be ready in the time it takes to boil milk. Make sure you have a food grade thermometer to ensure accuracy. The vinegar used in this recipe is simply to acidify the milk and will be discarded with the whey, so don't waste specialised flavored or imported vinegar. In this instance, cheap white vinegar is best because it has a high acidity level.

INGREDIENTS

5 litres (5 quarts) cow's milk (or goat's milk)
1 tablespooon salt
100 ml (3½ fl oz) plain white vinegar

METHOD

» Heat the milk to 60°C (140°F) over high heat, stirring constantly then add salt and stir well.
» Continue stirring while heating until the milk reaches 90°C (194°F), then as the milk edges past 90°C (194°F), add the vinegar and stir until flecks begin to appear on the surface. Remove from the heat and continue stirring for a few seconds until the liquid turns a clear almost greenish tinge and the coagulated curds are clearly visible.
» Allow to sit undisturbed for 5 minutes so that curds can rise to the surface.
» Using a slotted spoon, scoop curds into ricotta baskets or cheese moulds or even a colander and drain over the sink for 30 minutes. Transfer to a bowl or container and refrigerate until ready to use (up to 7 days).

FRESH HERB RICOTTA

Makes: Approximately 2 cups

Fresh ricotta is one of the most rewarding cheeses to make because it can literally be ready in the time it takes to boil milk. Make sure you have a food grade thermometer to ensure accuracy. The vinegar used in this recipe is simply to acidify the milk and will be discarded with the whey, so don't waste specialised flavored or imported vinegar. In this instance, cheap white vinegar is best because it has a high acidity level.

INGREDIENTS

5 litres (5 quarts) cow's milk (or goat's milk)
1 tablespoooon salt
100 ml (3½ fl oz) plain white vinegar
1 tablespoon chopped parsley and chives
1 teaspoon thyme or oregano
chilli flakes (optional)

METHOD

» Heat the milk to 60°C (140°F). over high heat, stirring constantly then add the salt and stir well. Continue stirring until the milk reaches 90°C/194°F (85°C/185°F for goat's milk). As the milk edges past 90°C (194°F), add all the herbs (and chilli, if using) and the vinegar. Stir until flecks of curd begin to appear on the surface. Remove from the heat and continue stirring for a few seconds until the liquid turns a clear almost greenish tinge and the coagulated curds are clearly visible.

» Allow to sit undisturbed for 5 minutes so that curds can rise to the surface.

» Using a slotted spoon, scoop curds into ricotta baskets or cheese moulds or even a colander and drain over the sink for 30 minutes. Transfer to a bowl or container and refrigerate until ready to use (up to 5 days).

WHEY RICOTTA

Makes: Approximately 2 cups

Fresh ricotta made from the whey—left over from cheese-making, is the original form or ricotta. This Italian word comes from the Italian for 're-cooked' because the leftover whey was reheated or cooked to force the remaining curds to the surface. In days of old, cheesemakers would heat this leftover whey and enjoy the resulting curds with bread for their lunch. The remaining whey was then fed to the pigs on the farm…nothing was ever wasted!

Because most of the proteins have already been removed from the whey in the prior cheesemaking venture, I suggest adding some whole milk to the whey to give you a worthwhile volume of ricotta.

INGREDIENTS

5 litres (5 quarts) fresh whey from cheesemaking (goat, cow, buffalo or sheep)
1–2 litres fresh full cream milk
1½ tablespooon salt
up to 4 tablespoons white vinegar, if necessary

METHOD

- Heat the milk and whey to 60°C (140°F) over high heat, stirring constantly, then add salt and stir well.
- Continue stirring while heating until the liquid reaches 90°C (194°F) then as the milk edges past 90°C (194°F), watch for the curds beginning to form. You may not need any vinegar as the whey is already acidified.
- If the temperature reaches 95°C (203°F) and no curds have formed, add up to 4 tablespoons white vinegar and stir until flecks begin to appear on the surface. Remove from the heat and continue stirring for a few seconds until the liquid turns a clear almost greenish tinge and the coagulated curds are clearly visible.
- Allow to sit undisturbed for 5 minutes so that curds can rise to the surface.
- Using a slotted spoon, scoop curds into ricotta baskets or cheese moulds or even a colander and drain over the sink for 30 minutes. Transfer to a bowl or container and eat while warm or refrigerate until ready to use (up to 7 days).

cow's milk cheese

WASHED RIND CHEESE (SOFT)

Makes: 4–6 medium cheeses

Aromatic, strong and not for the faint-hearted, washed rinds are quite divisive, but for lovers of strong flavors, this cheese is absolutely delicious…the smellier, the better!

INGREDIENTS

10 litres (10 quarts) cow's milk, preferably Jersey
pinch of MA4002
pinch of B Flora
pinch of Penicillium Candidum (PC)
pinch of Geotrichum 13 (GEO)
2 ml (0.006 fl oz) calcium chloride
2 ml (0.006 fl oz) rennet
salt
liquid for washing—beer, wine or a light brine (see Basics)

METHOD

» Heat the milk to 33°C (91°F), add the starter cultures, PC and GEO and stir well for 1 minute. Allow to rest for 30 minutes.

» Add the calcium chloride and stir well, then add the rennet and stir for a minimum of 1 minute and maximum of 2 minutes.

» Allow to sit undisturbed for 1 hour until the curd has set and when tested with a knife, you achieve a 'clean break' (see Basics – Techniques).

» Cut the curd into 2 cm (¾ in) squares, stir briefly to ensure no large pieces remain and allow to rest for 10 minutes (this is called 'healing'). If you do see any large pieces of curd, quickly cut these to the approximate size of the rest of the curd.

» Stir well for 2 minutes, then rest for a further 5 minutes.

RECIPE CONTINUES ON PAGE 48

- » Check the curd is beginning to coagulate by holding a handful of curd in your hand and gently make a fist. Hold the curd gently but firmly for 20 seconds, then open your fist. The majority of the curd should have begun melding together. If so, you are ready to proceed.
- » Whey off (remove excess whey from curd) almost down to the level of the curd, then scoop curds into several moulds of your choice and place on draining board for 1 hour. Turn the cheeses and return to the mould so that the bottom moulded sides are now on top. Leave for 1 hour then turn again. Turn again a few hours later. This ensures even drainage and a good shape (if curd is too fragile to turn initially, leave for an extra hour or two before turning).
- » The next day, brine in saturated brine (see Basics) for about 2 hours for 400 g (14 oz) brie – less for smaller cheese, longer for larger ones). Once the brining is complete, remove the cheese from the liquid and allow to drain on a wire rack or draining board overnight. The next day, turn the cheese and leave on the draining rack.
- » On day 4, wash the cheese with your chosen beer, wine or brine by gently brushing all over with a sanitised pastry or makeup brush dipped in the liquid, then transfer to sanitised ageing box or plastic container (see Basics – Step 13). Ideally the temperature should be 10–12°C (50–53°F) and humidity above 90%.
- » This is best achieved in a wine fridge or cool room such as a laundry or garage (in cool climates). If none of these options are available, the cheese can be 'ripened' in the fridge but it will take longer to develop a rind.
- » Twice a week, wash the cheese as per the first 'wash' then turn over and return to the container. Continue this process until the cheese is fragrant with a slightly sticky rind—about 2–3 weeks. Wrap the cheese in perforated cheese paper or baking paper and transfer to a small plastic container and store in a regular fridge for a further 1–2 weeks to ensure cheese has sufficiently ripened throughout. Cheese should be ready to eat 4 weeks after making but will continue to improve with age over the next 3–4 weeks. By 8 weeks of age, it may well be getting to strong and past its best.

cow's milk cheese

HALOUMI

Makes: Approximately 1 kg (2 lb 4 oz)

Traditionally in the Mediterranean, haloumi was made from sheep's milk but in Australia it tends to be made from cow's milk which makes it a little more bland. I love to use a combination of sheep or buffalo milk and cow's milk (50% of each) but you can use all sheep's milk or all buffalo milk if you wish.

INGREDIENTS

10 litres (10 quarts) sheep, buffalo or cow's milk or a blend
pinch of B Flora and/or MA4002
2 ml (0.006 fl oz) calcium chloride
2 ml (0.006 fl oz) vegetarian rennet

METHOD

- » Heat the milk to 33°C (91°F), add the cultures and stir well for 1 minute. Allow to rest for 30 minutes. Add the calcium chloride and stir well, then add the rennet and stir for a minimum of 1 minute and maximum of 2 minutes.
- » Allow to sit undisturbed for 1 hour until the curd has set and when tested with a knife, cuts cleanly without the curd slipping off the blade (see Basics – Techniques).
- » Cut the curd into 2 cm (¾ in) squares, stir briefly to ensure no large pieces remain. If you do see any large pieces of curd, quickly cut these to the approximate size of the rest of the curd. Allow the cut curds to rest for 10 minutes (this is called 'healing'). Stir well for 2 minutes, then rest for a further 5 minutes.
- » Gently increase the heat of the curds and whey to 37°C (98°F) over the next 15–20 minutes, cutting the curds again until they are about half the size they were. Stir well intermittently so that the curds do not clump together. Allow to rest for 5 minutes so that the curds settle and the whey is covering the surface. The curds should have shrunk and now appear visible smaller.
- » Place a wire rack over your sink. Line a large, sanitised rectangular draining basket with a piece of sanitised cheesecloth then scoop curds out of the whey and into the draining basket until all curds have been removed from the whey. Fold the remaining cheesecloth neatly over the curd then place on the wire rack over the sink to drain.

- » Place a clean flat tray over the cheesecloth and add some items weighing at least 2 kg (4 lb 8 oz). Cans of food work well. After 1 hour, remove the tray, turn over the cheese still wrapped in its cheesecloth and return to the basket. Replace the tray and weights and leave for 1 hour. Remove tray and weights, turn cheese again and return to the basket without the tray and weights. Leave overnight.
- » The next day, cut the large piece of curd into pieces weighing about 250 g (9 oz) each then place in a saturated brine (see Basics – Technique) for 2–3 hours, depending on how salty you want it to be. Remove from the brine and allow the cheese to drain on a clean, sanitised rack overnight.
- » The next day, place the cheese on a tray and chill until dry, turning once. About 12 hours each side.
- » Store in an airtight container for up to 1 week, or vacuum seal and store in the fridge for up to 6 months.

cow's milk cheese

TRADITIONAL MORBIER (FIRM WASHED RIND)

Makes: 1 large cheese

One of my favorite cheeses, this classic has a line of ash through the centre. In days gone by, this was done to deter insects from settling on the curds in the moulds while the cheesemaker waited for the evening milk to be brought in, turned into curd and filled the mould and complete the cheese. These days, the ash is still used to keep the old cheese-making traditions alive!

INGREDIENTS

10 litres (10 quarts) cow's milk
pinch of B Flora and/or MA4002
2 ml (0.006 fl oz) calcium chloride
2 ml (0.006 fl oz) vegetarian rennet
charcoal ash
bacterial brine for washing rind (brine with brevi linens added. See Basics)

METHOD

» Heat the milk to 33°C (91°F), add the cultures and stir well for 1 minute. Allow to rest for 30 minutes.

» Add the calcium chloride and stir well, then add the rennet and stir for a minimum of 1 minute and maximum of 2 minutes.

» Allow to sit undisturbed for 1 hour until the curd has set and when tested with a knife, cuts cleanly without the curd slipping off the blade (see Basics – Techniques).

» Cut the curd into 2 cm (¾ in) squares, stir briefly to ensure no large pieces remain and allow to rest for 10 minutes (this is called 'healing'). If you do see any large pieces of curd, quickly cut these to the approximate size of the rest of the curd. Stir well for 2 minutes, then rest for a further 5 minutes.

» Gently increase the heat of the curds and whey to 37°C (99°F) over the next 15–20 minutes, cutting the curds again until they are about half the size they were before (pea size) and stir well intermittently so that the curds do not clump together. Allow to rest for 5 minutes so that the curds settle and the whey is covering the surface.

RECIPE CONTINUES ON PAGE 54

- Remove whey until the liquid is level with the curds, measuring how much whey you discard, then replace this volume with an equal amount of water of the same temperature (37°C/99°F). Stir gently for the next 10 minutes, maintaining the heat of 37°C (99°F).

- Scoop the shrunken curds from the whey and divide evenly into your two cheese moulds that have been lined with cheesecloth. After 10 minutes, sprinkle the surface of the cheese with the ash, then gently remove one of the cheeses from its mould and place—ash side down—on the other cheese. Lift the sides of the cheesecloth up and tuck neatly over and around the top of the cheese and allow to rest for 30 minutes.

- After 30 minutes, you will need to press the cheese, place a sanitised tray onto the surface of the cheese. Press the sanitised 'press' onto the surface of the cheese and weight with a plate and a couple of cans of soup or other items weighing about 1 kg/2 lb 4 oz (make sure all items coming into contact with the cheese are clean).

- After 1 hour, remove the pressing items, remove the cheese from the mould and remove the cheesecloth. Return the cheesecloth to the mould and return the cheese, upside down then re-wrap cheesecloth as before and return pressing items and leave cheese overnight.

- The next day, unwrap the cheese and weigh. Multiply this weight by 3%. For example, if the cheese weighs 1 kg (2 lb 4 oz), you will need 30 g (1 oz) of salt.

- Holding the cheese carefully, rub the salt all over the top, bottom and sides of the cheese and then return to the mould without the cheesecloth. Allow to sit on a draining rack for 24 hours, turning once during that time.

- Meanwhile, make a bacterial brine by boiling water. To 200 ml (7 fl oz) of cooled boiled water, add 1 teaspoon salt and a pinch of 'Brevi Linens' (see Basics – Ingredients). Stir well.

- Prepare a deep container with a lid for the cheese, with a plastic draining rack lined with paper towel.

- Place the cheese on the rack, cover with the lid and place in a cool environment of about 12°C (54°F) such as a wine fridge or cool garage or laundry. If there is no suitable place, a fridge is fine but the cheese will take longer to mature.

- Every 3–5 days, remove the cheese, wipe out the container, replace the paper and 'wash' the cheese all over with the cool bacterial brine. I like to use a cleaned, sanitised makeup or pastry brush. Once the cheese is damp all over, return to its box, cover with the lid and return to the cool place. Continue this process for up to 2 months.

- After 2 months, the cheese should have a beige/grey aromatic rind. If you don't like the grey rind, brush gently with soft brush to remove then wash with a little bacterial brine and allow to sit in a well ventilated fridge for 2 days, turning after the first day.

- Wrap the cheese in baking paper or cheese paper and place in a plastic container in a cold fridge for up to 6 months. Note, the older the cheese, the stronger the flavor.

cow's milk cheese

FRENCH-STYLE BRIE

Makes: 4–6 medium cheeses

This is a typical, Northern French-style recipe that results in a brie that is soft, glossy and oozy with a mild, thin white mould rind.

INGREDIENTS

10 litres (10 quarts) cow's milk, preferably Jersey
pinch each of MA4002 and B Flora
pinch of Penicillium Candidum (PC)
pinch of Geotrichum 13 (GEO)
2 ml (0.006 fl oz) calcium chloride
2 ml (0.006 fl oz) rennet
salt

METHOD

» Heat the milk to 33°C (91°F) and add the starter cultures, PC and GEO and stir well for 1 minute. Allow to rest for 30 minutes.

» Add the calcium chloride and stir well, then add the rennet and stir for a minimum of 1 minute and maximum of 2 minutes.

» Allow to sit undisturbed for 1 hour until the curd has set and when tested with a knife, there is a clean 'break' (see Basics – Techniques).

» Cut the curd into 2 cm (¾ in) squares, stir briefly to ensure no large pieces remain and allow to rest for 10 minutes (this is called 'healing'). If you do see any large pieces of curd, quickly cut these to the approximate size of the rest of the curd.

» Stir well for 2 minutes, then rest for a further 5 minutes.

» Check the curd is beginning to coagulate by holding a handful of curd in your hand and gently make a fist. Hold the curd gently but firmly for 20 seconds then open your fist. The majority of the curd should have begun melding together. If so, you are ready to proceed.

RECIPE CONTINUES ON PAGE 57

- » Whey off (remove excess whey from curd) almost down to the level of the curd then scoop curds into the moulds of your choice and place on draining board for one hour. Turn the cheeses and return to the moulds so that the bottom moulded side is now on top. Leave for one hour then turn again. Turn again a few hours later. This ensures even drainage and a good shape (if curd is too fragile to turn initially, leave for an extra hour or two before turning).
- » The next day, brine in saturated brine (see Basics – Techniques) for about 2 hours for 200 g (7 oz) Brie—less for smaller cheese, longer for larger ones). Once the brining is complete, remove the cheese from the liquid and allow to drain on a wire rack or draining board overnight.
- » The next day, turn the cheeses and leave on draining rack.
- » On day 4, transfer to sanitised ageing box or plastic container. Ideally the temperature should be 10–12°C (50–53°F) and humidity above 90%.
- » This is best achieved in a wine fridge or cool room such as a laundry or garage (in cool climates). If none of these options are available, the cheese can be 'ripened' in the fridge but it will take longer to develop a rind (about 3 weeks).
- » Turn every 2 days to ensure even mould growth. Once the cheese is fully covered with an even white mould, wrap the cheese in perforated cheese paper or baking paper and transfer to a small plastic container and store in a regular fridge for a further 1–2 weeks to ensure cheese has sufficiently ripened throughout. Cheese should be ready to eat after 4 weeks.

Note: Your brie can be varied by the type of milk you use. Rich Jersey will give a golden brie, a regular milk will give a paler cheese. Choose a variety of moulds to shape your brie. I love square moulds but the round ones are more traditional.

Variation: If you love peppercorns, consider adding some crushed peppercorns to the curd after you have removed some of the whey but before you transfer the curds to their moulds. Add about 1 tablespoon of crushed black, pink and white peppercorns by sprinkling evenly over the curds, then mixing in with your hands.

cow's milk cheese

GREEK-STYLE FETA

Makes: Approximately 1 kg (2 lb 4 oz) feta

Mediterranean-style feta differs vastly between the various regions of the Mediterranean but I love the Greek style best. Traditionally made with sheep or goat's milk, these days it is often made with cow's milk which is easier to purchase and less expensive. If goat's milk is available, try using half cow and half goat for a more authentic flavor.

INGREDIENTS

10 litres (10 quarts) cow's milk or 5 litres (5 quarts) each cow's and goat's milk
large pinch mesophilic culture diluted in a little cooled boiled water
¼ teaspoon lipase
2 ml (0.006 fl oz) calcium chloride
2 ml (0.006 fl oz) vegetarian rennet
salt

METHOD

» Heat the milk to 30°C (86°F), adding the culture and lipase to the warming milk. Stir gently to distribute. Allow the milk to sit for a minimum 30 minutes and no more than 1 hour. Add the calcium chloride and stir gently. Add the rennet and stir gently for 1 minute, then allow the milk to set, undisturbed for 1 hour.

» Once the milk has set, make a single cut to check for a 'clean break' (see Basics – Techniques) then cut curds into 1 cm (½ in) cubes. (If a 'clean break' is not achieved, leave milk to set for a further 15 minutes then proceed.)

» Stir curds gently for 15 minutes until they have begun to shrink and whey is accumulating. Allow to 'rest' for 5 minutes until the entire surface of the vat is whey and the curds have sunk to the bottom.

RECIPE CONTINUES ON PAGE 60

- » Scoop the curds into a cheesecloth-lined colander then gather the sides of the cheesecloth to enclose the curds. Tie the cheesecloth over a wooden spoon, then hang the curds over a sink to drain for 4 hours. Alternatively, scoop the curds into square moulds so that you have even blocks of cheese. Either method is fine.
- » If you choose to use square moulds, fill the moulds to full and allow the curd to coagulate. After 30 minutes, remove the cheese from the mould, turn the cheese and return to the mould so that the top is now on the bottom. Turn again after 1 hour.
- » The following day, you need to add salt. Traditionally this is done by sprinkling salt on the sliced cheese. Before you start, weigh all the cheese and then multiply the weight by 4%. For example, if you have 1 kg (2 lb 4 oz) of cheese, you will need 40 g (1½ oz) of salt.
- » Cut the cheese into thick 3 cm (1¼ in) slices, then lie flat on a draining rack. Sprinkle evenly with the required amount of salt, using half the salt initially, turning the cheese and sprinkling the remainder over the cheese. Allow to drain on the rack for 24 hours, turning once during that time.
- » The cheese is now ready to eat and should be consumed within a week, stored in the refrigerator and packed in a plastic container with a lid.
- » If you prefer to preserve the cheese for up to 3 months, make a simple light brine by mixing 1 litre (34 fl oz) of boiling water with 100 g (3½ oz) salt and stir well. Cool thoroughly.
- » When the brine is cool, transfer to a suitable container that can take the brine and the cheese and pack the cheese neatly into the container. Pour the brine over. The cheese should be submerged under the brine and kept in the refrigerator for up to 3 months.

cow's milk cheese

STILTON-STYLE BLUE CHEESE

Makes: 1 large cheese

Typically English, this cow's milk cheese is not difficult to make but does take practice to get the texture right. With a long maturation process, this cheese is always great to have on hand.

INGREDIENTS

10 litres (10 quarts) cow's milk
pinch of Flora Danica or B Flora
pinch of MA4001
pinch of Roquefort (blue mould) spores
2 ml (0.006 fl oz) calcium chloride
2 ml (0.006 fl oz) rennet
salt, to taste

METHOD

» Heat the milk to 33°C (91°F) and during the heating, add the cultures. Blue mould can be added now or later when the curd is set. Stir gently to dissolve and distribute the cultures. Allow the milk to rest for 30 minutes.

» Add calcium chloride, stir then add the rennet and stir for 90 seconds, then allow to set.

» After 1 hour, check the curd for a 'clean break' (Basics – Techniques). Cut the curd evenly into 2 cm (¾ in) cubes, then stir gently and rest for 5 minutes. Cut the curd again and stir then allow the curds to sink, about 10 minutes. Whey off as much as possible. Add blue mould spores now if they weren't added previously.

» Transfer the curd to cheesecloth-lined colander and allow to rest for a few minutes, then break up with your hands. Add salt to taste (I like 2 tablespoons per 10 litres/10 quarts) and mix the salt through the curd, breaking up any large pieces as you go.

RECIPE CONTINUES ON PAGE 63

- When the curd feels dry and crumbly to touch, add the curd to sanitised moulds by dropping (not packing) the curd into the mould. We want to keep air holes amongst the curd. Overfill the mould to allow for sinkage. Allow the curd to knit together, then turn carefully after 30 minutes. Turn again once more that day.
- The following day (day 2) weigh the cheese and add 1% salt and rub over the exterior of the cheese. Return the cheese to the mould and allow to sit on a draining mat for a further day, turning once or twice.
- The following day (day 3), remove the cheese from its mould and allow to sit on the draining mat for a further day.
- On day 4, the cheese should be dry and ready to transfer to a plastic aging box with a draining mat and lid. Maintain a temperature of about 10–14°C (50–57°F) and relatively high humidity of about 85% by using a lid on the box but not sealed.
- Turn the cheese every few days. When blue mould begins to form, piece the cheese with a sanitised knitting needle or stainless steel rod and return to the aging box. Check the temperature and humidity and turn every week or so (more often if you are diligent).
- Brush off any moisture or unsightly mould that appears on the surface, or leave until cheese is mature—2 months for small cheese, 4–5 months for larger ones.
- When the cheese smells like a blue cheese and you think it is ready, cut in half and taste. If you are happy with the flavor and texture, wrap each half in cheese paper or baking paper and place in a plastic container in your refrigerator. Enjoy over the next 4–6 months.

SOFT BLUE GOAT CHEESE

Makes: 8–10 small round cheeses

This rennet set goat cheese can be soft and oozy or firmer, depending on how you cut the curd. Blue mould will only grow on the surface so pierce for internal blue or leave for a blue rind only.

INGREDIENTS

10 litres (10 quarts) goat's milk
pinch of mesophilic culture
pinch of blue mould
pinch of lipase (optional) diluted in 2 tablespoons cooled, boiled water
2 ml (0.006 fl oz) calcium chloride
2 ml (0.006 fl oz) rennet
salt at 2%, or to taste
ash charcoal powder, optional

METHOD

» Heat the milk to 32°C (90°F), adding cultures and blue mould as milk is gently warmed. Blue mould can be added after whey is removed if you prefer, but I find I get a better result if the blue mould spends more time in the warm milk bath so I prefer to add now.

» Add the calcium chloride and lipase and stir thoroughly. Add the rennet and stir for 90 seconds, then allow to set for 1 hour.

» Make a single cut and check for 'clean break' (see Basics – Techniques). Cut gently into 2 cm (¾ in) cubes, then allow to rest for 5 minutes. Stir gently then allow curd to 'heal' for a few minutes. 'Whey Off' (see Basics – Techniques). Add blue mould now if not added previously.

» Fill the moulds and continue topping up as the whey drains. Once the moulds are full, allow the cheese to drain for 30 minutes, then turn.

RECIPE CONTINUES ON PAGE 66

- » If you prefer a firmer blue goat cheese, stack the cheese 'two high' so that each cheese sits on another. After 1 hour, turn each cheese in its mould and place the upper cheeses underneath the other cheese so that the bottom cheese is pressed. After a further hour, turn all the cheeses in their moulds and unstack so that each is sitting on a draining rack.
- » The following day, (day 2) weigh each cheese and measure 2% salt, for example if a cheese weights 200 g (7 oz) you will require 4 g (⅛ oz) salt for that cheese. Rub gently on all outer areas, then return to its mould for a further day.
- » The following day, (day 3) sanitise a knitting needle or skewer and pierce the cheese about 10 times, through the top. Take care not to break the cheese. If you prefer the blue mould growing only on the surface, do not pierce the cheese.
- » The humidity and temperature of a cheese is crucial for blue mould growth. The cheese must not be wet or damp when transferring to an aging box, or the blue mould will not grow.
- » On day 3, you should see the first tiny signs of blue mould. Place the cheese in its aging box (see Basics – Equipment) without the lid and leave at cool room temperature, such as in the kitchen, for two more days (turning the cheese each day) to get the blue mould growing. Once well covered, place the lid on the tub without completely sealing it and transfer to a fridge or bar fridge. This cheese is ideally aged at 11–15°C (51–59°F) but if your only option is the fridge, it will simply take longer to ripen.
- » If you want to 'ash' the cheese, you will now dust it lightly with the powdered ash or charcoal. Over the next few days, the blue mould and natural white mould will grow over and cover the ash, but it will be visible when you cut the cheese.
- » Every 3–4 days, remove the cheese box, remove the cheese and wipe out the container to get rid of excess condensation. Return the cheese on its plastic mat on fresh paper towel and return to the fridge, with the lid on or ajar as you prefer. Leaving the lid ajar will produce a harder blue cheese.
- » After 2–3 weeks in a wine fridge or 3–4 weeks in a domestic fridge, your cheese should be ready. Wrap in cheese paper or baking paper and place in plastic container until you are ready to eat.

goat's milk cheese

FRENCH GOAT CROTTIN

Makes: 4–6 medium cheeses

These delicious French-style goat cheeses have their own different characteristics, depending on where in France you buy them and how old they are. Some are firm and crumbly with blue spots, some are soft and oozy and others are so hard you can only grate them… all are delicious! This recipe produces my favorite—a cheese that has a firm centre but is oozy just under the rind. If you can manage to keep it for about 6–8 weeks and you love strong cheese, this one will just about run out of its rind when cut!

INGREDIENTS

10 litres (10 quarts) goat's milk
pinch of MA4002
pinch of B Flora
pinch of Penicillium Candidum (PC)
pinch of Geotrichum 17 (GEO)
2 ml (0.006 fl oz) calcium chloride
2 ml (0.006 fl oz) rennet
salt

METHOD

» Heat the milk to 33°C (91°F) and add the starter cultures, PC and GEO and stir well for 1 minute. Allow to rest for 30 minutes.

» Add the calcium chloride and stir well then add the rennet and stir for a minimum of 1 minute and maximum of 2 minutes.

» Allow to sit undisturbed for 1 hour until the curd has set and when tested with a knife, there is a clean 'break' (see Basics – Techniques).

» Cut the curd into 2 cm (¾ in) squares, stir briefly to ensure no large pieces remain and allow to rest for 10 minutes (this is called 'healing'). If you do see any large pieces of curd, quickly cut these to the approximate size of the rest of the curd.

RECIPE CONTINUES ON PAGE 70

- Stir well for 2 minutes, then rest for a further 5 minutes.
- Check the curd is beginning to coagulate by holding a handful of curd in your hand and gently make a fist. Hold the curd gently but firmly for 20 seconds then open your fist. The majority of the curd should have begun melding together. If so, you are ready to proceed. If not, wait a few more minutes before testing again.
- Whey off (remove excess whey from curd) almost down to the level of the curds, then scoop the curds into round cheese mould of your choice, making sure you completely fill the moulds. Place the new cheeses on draining rack for 1 hour, then carefully remove the cheese from the mould, turn and return to the mould. After another hour, turn the cheeses again.
- Turn again 2 hours later. This ensures even drainage and a good shape (if curd is too fragile to turn initially, leave for an extra hour or 2 before turning).
- The next day, brine in saturated brine (see Basics – Techniques) for about 2 hours for 400 g (14 oz) cheese—less for smaller cheese, longer for larger ones). Once the brining is complete, remove the cheese from the liquid and allow to drain on a wire rack or draining board overnight.
- The next day, turn the cheese and leave on the draining rack.
- On day 4, transfer to sanitised ageing box or plastic container. Ideally the temperature should be 10–12°C (50–53°F) and humidity above 90%.
- This is best achieved in a wine fridge or cool room such as a laundry or garage (in cool climates). If none of these options are available, the cheese can be 'ripened' in the fridge but it will take longer to develop a rind.
- Turn every 2 days to ensure even mould growth, wiping out the box if it is wet. Once the cheese is fully covered with an even white mould, (2–3 weeks) wrap the cheese in perforated cheese paper or baking paper and transfer to a small plastic container and store in a regular fridge for a further 2–4 weeks to ensure cheese has sufficiently ripened throughout. Cheese should be ready to eat in 3 weeks and will begin to deteriorate after about 8 weeks.

Note: If you prefer a harder goat cheese, do not cover them in their aging box after day 5, but leave the lid ajar to allow airflow. Follow all other steps as above.

goat's milk cheese

PURE GOAT BRIE

Makes: 4–6 medium cheeses

This is a typical, Northern French style recipe that results in a brie that is soft, glossy and oozy with a mild, thin white mould rind. While it isn't typical to use goat's milk to make brie in France, my customers and students love this cheese and its mild interior which is less rich than a cow's milk brie.

INGREDIENTS
10 litres (10 quarts) goat's milk
pinch of MA4002
pinch of B Flora
pinch of Penicillium Candidum (PC)
pinch of Geotrichum 13 (GEO)
2 ml (0.006 fl oz) calcium chloride
2 ml (0.006 fl oz) rennet
salt

METHOD
- Heat the milk to 33°C (91°F) and add the starter cultures, PC and GEO and stir well for 1 minute. Allow to rest for 30 minutes.
- Add the calcium chloride and stir well, then add the rennet and stir for a minimum of 1 minute and maximum of 2 minutes.
- Allow to sit undisturbed for 1 hour until the curd has set and when tested with a knife, there is a clean 'break' (see Basics – Techniques).
- Cut the curd into 2 cm (¾ in) squares, stir briefly to ensure no large pieces remain and allow to rest for 10 minutes (this is called 'healing'). If you do see any large pieces of curd, quickly cut these to the approximate size of the rest of the curd.
- Stir well for 2 minutes, then rest for a further 5 minutes.

RECIPE CONTINUES ON PAGE 73

- » Check the curd is beginning to coagulate by holding a handful of curd in your hand and gently make a fist. Hold the curd gently but firmly for 20 seconds, then open your fist. The majority of the curd should have begun melding together. If so, you are ready to proceed. If not, wait a few more minutes before testing again.
- » Whey off (remove excess whey from curd) almost down to the level of the curd, then scoop curds into a mould of your choice and place on draining board for 1 hour. Turn the cheese and return to the mould so that the bottom moulded side is now on top. Leave for 1 hour then turn again. Turn again a few hours later. This ensures even drainage and a good shape (if curd is too fragile to turn initially, leave for an extra hour or two before turning).
- » The next day, brine in saturated brine (see Basics – Techniques) for about 2 hours for 400 g (14 oz) brie—less for smaller cheese, longer for larger ones). Once the brining is complete, remove the cheese from the liquid and allow to drain on a wire rack or draining board overnight. The next day, turn the cheese and leave on draining rack. On day 4, transfer to sanitised ageing box or plastic container. Ideally the temperature should be 10–12°C (50–53°F) and humidity above 90%.
- » This is best achieved in a wine fridge or cool room such as a laundry or garage (in cool climates). If none of these options are available, the cheese can be 'ripened' in the fridge but it will take longer to develop a rind.
- » Turn every 2 days to ensure even mould growth. Once the cheese is fully covered with an even white mould, wrap the cheese in perforated cheese paper or baking paper and transfer to a small plastic container. Store in a regular fridge for a further 1–2 weeks to ensure cheese has sufficiently ripened throughout. Cheese should be ready to eat in 3 weeks and will begin to deteriorate after about 8 weeks.

FRESH HERBED GOAT CHEESE

Makes: Approximately 10 small cheese rounds

I love the flavor of these small, French-style herbed goat cheese rounds. You can roll them in any fresh chopped herbs from your garden, or finely chopped nuts, paprika, chilli, crushed dried tomato or any other flavoring you prefer.

INGREDIENTS

5 litres (5 quarts) goat's milk
pinch of mesophilic culture
1 ml (0.03 fl oz) calcium chloride
0.5 ml (0.015 fl oz) rennet
salt

METHOD

» Heat the milk to 25°C (77°F), add the culture and stir well for 1 minute. Allow to rest for 30 minutes. Add the calcium chloride and stir well, then add the rennet and stir for a minimum of 1 minute and maximum of 2 minutes.

» Allow to sit undisturbed for 14–24 hours until the curd has set. Note the curd will be very soft and fragile – much like yoghurt. The longer you leave it, the stronger the goat flavor.

» Transfer the curd to a cheesecloth-lined colander, cover the surface of the curd with the overhanging cheesecloth and allow to drain for 2–4 hours, depending on how thick you want your curd.

» Add salt to taste (I add 1 tablespoon salt for this recipe). Mix the salt into the curd thoroughly, taking care to mix the firmer curd down the bottom of the colander. The texture should be smooth.

» Spoon the salted curd into six small, sanitised cheese moulds and allow these to sit on a drainer or rack for 24 hours. The next day, carefully remove cheese from the moulds, turn over then return them to the moulds and allow to drain for a further 24 hours.

» Transfer in their moulds to the fridge and chill for 24 hours. They will still drip a little so place on a tray to avoid mess in the fridge.

» Once chilled, the cheeses can be rolled in your choice of fresh or dry herb and spices.

» Cheeses should be stored in a covered container in the fridge and consumed within 3 weeks.

goat's milk cheese

PERSIAN (MARINATED) FETA

Makes: Approximately 1 kg (2 lb 4 oz)

Persian feta is one of the most popular cheeses at our farmers markets. The lovely, light fresh goat cheese is marinated in oil that can be further enhanced with herbs or spices for extra flavor. Make sure that once marinated, all surfaces of the cheese are under the oil to create an oxygen barrier and prevent spoilage. Two straight-sided rectangular colanders makes the shaping of the cheese easier.

INGREDIENTS

5 litres (5 quarts) goat's milk
pinch of mesophilic culture
1 ml (0.03 fl oz) calcium chloride
0.5 ml (0.015 fl oz) rennet
salt

METHOD

- » Heat the milk to 25°C (77°F), add the cultures and stir well for 1 minute. Allow to rest for 30 minutes. Add the calcium chloride and stir well, then add the rennet and stir for a minimum of 1 minute and maximum of 2 minutes.
- » Allow to sit undisturbed for 20–24 hours until the curd has set. Note the curd will be very soft and fragile—much like yoghurt.
- » Transfer the curd to a cheesecloth-lined colander, cover the surface of the curd with the overhanging cheesecloth and allow to drain for 6 hours. Transfer to a mixing bowl and add salt to taste (as a guide, I add 1 tablespoon of salt for this recipe).
- » Divide the salted curd between two cheesecloth-lined square or rectangular drainage baskets or colanders, preferably with straight sides. Use the overhanging fabric to cover the curd, then stack one colander neatly over the other to weight it. Allow these to drain on a rack over the sink for 12 hours, then swap the top colander to the bottom to weight the other cheese. Drain for a further 12 hours.
- » Transfer both colanders to a tray and chill in the fridge for a minimum of 24 hours until firm enough to cut.
- » Cut the cheese into square or rectangular pieces and pack into glass jars or containers and pour high-quality olive oil over the surface, taking care that all cheese is submerged. If you are adding spices or herbs, you can add to the oil before covering the cheese or you can add after filling. These should also be submerged unless they are dried. Use within 3 months.

FRESH GOAT CURD

Makes: Approximately 1 litre (35 fl oz)

This wonderfully fresh goat curd is a light, fresh and spreadable goat cheese. It is also low in fat which is a bonus. As this cheese is slow to set, allow a day or two to complete the process. For a mild flavor, set the curd for 14 hours. For a deeper and more rustic flavor, allow the milk to set for 24 hours.

INGREDIENTS

3 litres (3 quarts) goat's milk
pinch of mesophilic culture
0.5 ml (0.015 fl oz) calcium chloride
0.5 ml (0.015 fl oz) rennet
salt

METHOD

- Heat the milk to 25°C (77°F) and add the cultures and stir well for 1 minute. Allow to rest for 30 minutes. Add the calcium chloride and stir well, then add the rennet and stir for a minimum of 1 minute and maximum of 2 minutes.
- Allow to sit undisturbed for 14–24 hours until the curd has set. Note the curd will be very soft and fragile—much like yoghurt.
- Transfer the curd to a cheesecloth-lined colander, cover the surface of the curd with the overhanging cheesecloth and allow to drain for 2–4 hours, depending on how thick you want your curd.
- Add salt to taste (I add 2–3 teaspoons salt for this recipe). Mix the salt into the curd thoroughly, taking care to mix the firmer curd at the bottom of the colander. The texture should be smooth.
- If you prefer a firmer texture, allow the curd to continue to drain for a further 2–4 hours then mix until smooth. Transfer to a non-reactive container such as glass or into a selection of glass jars and chill.
- Use within 4 weeks.

SEMI-HARD GOAT CHEESE WITH PEPPERCORNS

Makes: 3–4 medium cheeses

I love the firm, dense texture of this cheese which develops a firm rind and is studded with crushed peppercorns. You could use chilli or other herbs or spices instead of the peppercorns if you prefer.

INGREDIENTS
10 litres (10 quarts) goat's milk
pinch of MA4002
pinch of B Flora
pinch of Penicillium Candidum (PC)
pinch of Geotrichum 13 (GEO)
2 ml (0.006 fl oz) calcium chloride
2 ml (0.006 fl oz) rennet
2 tablespoons crushed peppercorns (not ground pepper)
salt

METHOD
- Heat the milk to 33°C (91°F) and add the starter cultures, PC and GEO and stir well for 1 minute. Allow to rest for 30 minutes.
- Add the calcium chloride and stir well, then add the rennet and stir for a minimum of 1 minute and maximum of 2 minutes.
- Allow to sit undisturbed for 1 hour until the curd has set and when tested with a knife, there is a clean 'break' (see Basics – Techniques).
- Cut the curd into 2 cm (¾ in) squares, stir briefly to ensure no large pieces remain and allow to rest for 10 minutes (this is called 'healing'). If you do see any large pieces of curd, quickly cut these to the approximate size of the rest of the curd.
- Stir well for 2 minutes, then rest for a further 5 minutes.
- Using your cheese cutter or knife, again cut the curds so that the pieces of curd are now 'pea size', show signs of shrinkage and firmer than previously. Rest for 5 minutes.

RECIPE CONTINUES ON PAGE 82

- » Check the curd is beginning to coagulate by holding a handful of curd in your hand and gently make a fist. Hold the curd gently but firmly for 20 seconds, then open your fist. The majority of the curd should have begun melding together. If so, you are ready to proceed. If not, wait a few more minutes before testing again.
- » Whey off (remove excess whey from curd) almost down to the level of the curd, then sprinkle the crushed peppercorn over the curd. Stir for 5 minutes to firm up the curd and evenly distribute the peppercorns. Scoop curds into a mould of your choice and place on draining board, stacking cheeses 'two high' so that one on top is being used as a weight for the lower cheese. After 30 minutes, swap the bottom cheese to the top.
- » After an hour, unstack cheese and turn them all by removing the cheese from its mould, turning it over and returning to the mould so that the bottom moulded side is now on top. Leave for 1 hour, then turn again. Turn again a few hours later. This ensures even drainage and a good.
- » The next day, brine in saturated brine (see Basics – Techniques) for about 4 hours for 500 g (1 lb 2 oz) cheese (less for smaller cheese, longer for larger ones). Once the brining is complete, remove the cheese from the liquid and allow to drain on a wire rack overnight. The next day, turn the cheese and leave on draining rack. If you prefer, you can 'dry salt' the cheese. (See Basics – Techniques.)
- » The following day, transfer to sanitised ageing box or plastic container with a lid.
- » For 'ripening the cheese' (see Basics – Techniques), the ideal temperature should be 10–12°C (50–53°F) and humidity above 90%. This is best achieved in a wine fridge or cool room such as a laundry or garage (in cool climates). If none of these options are available, the cheese can be 'ripened' in the fridge but it will take longer to develop a rind.
- » Because we want this to be a firm cheese with a dry rind, we will ripen in the aging box but we won't seal with a lid. The lid should be placed across the box so that air can circulate but the cheese is protected enough that it won't dry out.
- » Turn every 5 days to stop the cheese sticking to the cheese mat. After 4 weeks, the cheese should be covered with a dry, slightly powdery rind. I like to rub it back gently with my fingers. Wrap the cheese in perforated cheese paper or baking paper. Transfer to a small plastic container and store in a regular fridge for a further 1–2 weeks to ensure the cheese has sufficiently ripened throughout. Cheese should be ready to eat 6 weeks from making and should be good to eat for up to 4 months.

goat's milk cheese

ASHED GOAT LOGS

Makes: Approximately 4–6 logs

Ashed goat logs are very typical of South and Central France. The fresh goat's milk is inoculated with white mould, which then begins to grow through the ash on about day 4–5.
The cheese is ready to eat after about day 3 as a fresh cheese or can be matured until the rind grows and the cheese interior begins to become oozy (from 2 to 4 weeks).

INGREDIENTS

5 litres (5 quarts) goat's milk
pinch of mesophilic culture
pinch of white mould (Penicillium Candidum)
1 ml (0.03 fl oz) calcium chloride
1 ml (0.03 fl oz) rennet
1 tablespoon salt or to taste
vegetable ash, for cheese making

METHOD

- Heat the goat's milk to 25°C (77°F), then add the culture and white mould and stir well for 1 minute. Allow to rest for 30 minutes. Add the calcium chloride and stir well, then add the rennet and stir for at least 1 minute but no more than 2 minutes.
- Allow to sit undisturbed for 14–24 hours until the curd has set. Note the curd will be very soft and fragile—much like yoghurt. The longer you leave it, the stronger the goat flavor.
- Transfer the curd to a sanitised cheesecloth-lined colander using a large jug or ladle, then cover the surface of the curd with the overhanging cheesecloth. Allow to drain for 2–4 hours, depending on how thick you want your curd.
- Add salt to taste. Mix the salt into the curd thoroughly, taking care to mix the firmer curd at the bottom of the colander. The texture should be smooth.

RECIPE CONTINUES ON PAGE 84

» Spoon the salted curd into six log-shaped, sanitised cheese moulds and place on a drainer or rack for 24 hours. The next day, carefully remove the cheese from the moulds, turn over and return the cheese to the moulds upside down and allow to drain for a further 24 hours.

» Transfer the moulds to the fridge and chill for 24 hours. They will still drip a little so place on a tray to avoid mess in the fridge.

» Once chilled, gently remove the cheeses from their moulds and sprinkle with the vegetable ash. (Please note, this is a messy job so I use a tray lined with paper towel.) Make sure the cheeses are well covered.

» Transfer the ashed cheese to a sanitised ageing box or plastic container with a well-fitting lid. Ideally the temperature should be 10–12°C (50–53°F) and humidity above 90%. This is best achieved in a wine fridge or cool room such as a laundry or garage (in cool climates). If none of these options are available, the cheese can be 'ripened' in the fridge but it will take longer to develop a rind.

» Turn every two days to ensure even mould growth, wiping out the box if it is wet. Once the cheese is fully covered with an even mould growth (2–3 weeks), gently pat the surface of the cheese to achieve an even black and white rind. Wrap the cheese in perforated cheese paper or baking paper and transfer to a small plastic container. Store in a regular fridge and eat now, or store for a further 2–4 weeks to ensure cheese has sufficiently ripened throughout. This cheese is ready to eat anytime after it is ashed, as a fresh cheese but will take 2–3 weeks for it to be fully developed with a rind. It will still be firm in the centre after this time and after another week or two, will be oozy and softening and stronger in flavor.

BUFFALO BRIE

Makes: 4–6 medium cheeses

I love the richness of this stark, white brie with a deliciously golden oozy texture and light, white mould rind. You can add crushed peppercorns to the curd before hooping, for an interesting twist!

INGREDIENTS

10 litres (10 quarts) buffalo milk
pinch of MA4002
pinch of B Flora
pinch of Penicillium Candidum (PC)
pinch of Geotrichum 13 (GEO)
2 ml (0.006 fl oz) calcium chloride
2 ml (0.006 fl oz) rennet
salt

METHOD

» Heat the milk to 33°C (91°F) and add the starter cultures, PC and GEO and stir well for 1 minute. Allow to rest for 30 minutes.

» Add the calcium chloride and stir well. Add the rennet and stir for a minimum of 1 minute and maximum of 2 minutes.

» Allow to sit undisturbed for 1 hour until the curd has set and when tested with a knife, there is a clean 'break' (see Basics – Techniques).

» Cut the curd into approx. 2 cm (¾ in) squares. Stir briefly to ensure no large pieces remain, cutting any that are visible and allow to rest for 10 minutes (this is called 'healing').

» Stir well for 2 minutes, then rest for a further 5 minutes.

» Check the curd is beginning to coagulate by holding a handful of curd in your hand and gently make a fist. Hold the curd gently but firmly for 20 seconds, then open your fist. The majority of the curd should have begun melding together. If so, you are ready to proceed. If you are adding peppercorns or other flavors, do so now.

» Whey off (see Basics – Technique) almost down to the level of the curd, then scoop curds into moulds of your choice and place on draining rack over sink for 1 hour. Turn each cheese and return

to the mould so that the bottom moulded side is now on top. Leave for an hour then turn again. Turn again a few hours later. This ensures even drainage and a good shape (if curd is too fragile to turn initially, leave for an extra hour or two before turning).

» The next day, brine in saturated brine (see Basics – Techniques) for about 2 hours for 200 g (7 oz) Brie—less for smaller cheese, longer for larger ones). Once the brining is complete, remove the cheese from the liquid and allow to drain on a wire rack or draining board overnight. The next day, turn the cheese and leave on draining rack. On day 4, transfer to sanitised ageing box or plastic container with a well fitting lid. Ideally the temperature should be 10–12°C (50–54°F) and humidity above 90%. This is best achieved in a wine fridge or cool room such as a laundry or garage (in cool climates). If none of these options are available, the cheese can be 'ripened' in the fridge but it will take longer to develop a rind (see Basics – Technique).

» Turn every 2 days to ensure even mould growth. Once the cheese is fully covered with an even white mould, wrap the cheese in perforated cheese paper or baking paper. Transfer to a small plastic container and store in a regular fridge for a further 1–2 weeks to ensure cheese has sufficiently ripened throughout. Cheese should be ready to eat after 4 weeks.

LACTIC BUFFALO CHEESE

Makes: Approximately 6 small cheese

Both buffalo and sheep's milk are high in protein and are fabulous for cheesemaking. In this recipe, the slow lactic style results in a deliciously fresh, light cheese that can be flavored with herbs, chilli, crushed dried tomatoes, etc.

INGREDIENTS
3 litres (3 quarts) buffalo or sheep milk
pinch of mesophilic culture
0.5 ml (0.015 fl oz) calcium chloride
0.5 ml (0.015 fl oz) rennet
salt

METHOD
» Heat the milk to 25°C (77°F), add the culture and stir well for 1 minute. Allow to rest for 30 minutes. Add the calcium chloride and stir well, then add the rennet and stir for a minimum of 1 minute and maximum of 2 minutes.
» Allow to sit undisturbed for 14–24 hours until the curd has set. Note the curd will be very soft and fragile, much like yoghurt. The longer you leave it, the stronger the flavor.
» Using a large ladle, transfer the curd to a cheesecloth-lined colander then encase the curd with the overhanging cheesecloth and allow to drain for 4–6 hours.
» Add salt to taste (I add 2 teaspoons salt for this recipe). Mix the salt into the curd thoroughly, taking care to mix the firmer curd at the bottom of the colander. The texture should be smooth. At this stage, you can add herbs, spices or chilli to flavor the curd, if desired.
» Carefully spoon the salted curd into six small, sanitised cheese moulds and allow these to sit on a drainer or rack for 24 hours. The next day, carefully remove cheese from the moulds, turn over then return them to the moulds and allow to drain for a further 24 hours.
» Transfer in their moulds to the fridge and chill for 24 hours. They will still drip a little so place on a tray to avoid mess in the fridge.
» Once chilled, the cheeses can be rolled in your choice of fresh or dry herb and spices.
» Cheeses should be stored in a covered container in the fridge and consumed within 3 weeks.

buffalo and sheep's and mixed milk cheese

LACTIC BLUE BUFFALO CHEESE

Makes: Approximately 4–6 cheeses

Buffalo milk is a wonderfully rich milk which makes fabulous surface ripened cheese. Most Buffalo milk in Australia is used for Mozzarella, which I think is a shame. Try this easy, delicious cheese which is ready to eat in a week and I think you will agree! As this cheese is slow to set, allow a day or two to complete the process. For a mild flavor, set the curd for 14 hours. For a deeper and more rustic flavor, allow the milk to set for 24 hours.

INGREDIENTS

5 litres (5 quarts) buffalo milk
pinch of mesophilic culture
pinch of blue mould spores
1 ml (0.03 fl oz) rennet
salt, to taste

METHOD

» Heat the milk to 25°C (77°F). Add the culture and blue mould and stir well for 1 minute. Allow to rest for 1 hour.
» Add the rennet and stir for a minimum of 1 minute and maximum of 2 minutes. Allow to sit undisturbed for 14–24 hours until the curd has set. Note the curd will be very soft and fragile—much like yoghurt.
» Transfer the curd to a cheesecloth-lined colander, cover the surface of the curd with the overhanging cheesecloth and allow to drain for 4 hours.
» Add salt to taste (I add 1 tablespoon of salt for this recipe but you can add more or less as you prefer). Mix the salt into the curd thoroughly, taking care to mix in the firmer curd at the bottom of the colander. Allow the salted curd to drain for a further 1–2 hours, then divide the mixture evenly into 6–8 small cheese moulds. The curd will drain and shrink to about half its size so make sure you fill the moulds completely. Allow to drain overnight.

RECIPE CONTINUES ON PAGE 91

- » The following day, very carefully turn the fragile cheeses and return to their moulds and continue to drain for a further 24 hours.
- » On day 3, remove the cheeses from their moulds and place in a sanitised aging box (see Basics – Equipment) on a plastic cheese mat over paper towel and leave for a further 24 hours, uncovered.
- » On day 4, you should start to see some specks of blue mould. Turn the cheeses and leave for another day.
- » On day 5, place the lid ajar on the box and transfer the cheese to a cool place at about 12°C (54°F). A wine fridge works well. If you don't have access to a wine fridge, a garage or unheated laundry in the cooler months works well. Otherwise transfer to a regular fridge but place the lid on properly to avoid the cheese drying out.
- » Each 2–3 days, remove the cheeses from their box, wipe out the box and replace the cheese on fresh paper towel under the draining mat. Cover and return to the cool space.
- » This cheese is ready to eat once fully covered with blue mould. After about a week, you may notice the blue mould being covered with a fine white mould. This is normal. If you prefer the blue, gently rub the outer surface with your fingertips to bring out the blue color again.
- » After two weeks, wrap the cheese in breathable cheese paper or baking paper and store in the fridge in a sealed plastic container for up to 8 weeks. Note the cheese will be firm and mild for the first 3 weeks, beginning to soften under the rind in weeks 3–5 and be very soft and stronger in flavor after week 6. It can be eaten at any stage of its life.

buffalo and sheep's and mixed milk cheese

HOMEMADE BUFFALO MOZZARELLA

Makes: 8–10 medium balls

There is something very special about making and stretching your own mozzarella at home… Not only do you have it available and fresh whenever you want it, but the flavor of a very fresh mozzarella is so much more delicious than one that is days old…Note that this recipe uses citric acid rather than cultures to acidify the milk, making it even faster to make!

INGREDIENTS

4 litres (135 fl oz) buffalo milk
1½ teaspoons citric acid
1 ml (0.03 fl oz) liquid rennet
300 ml (10 fl oz) water
1–2 teaspoons salt

METHOD

» Place the milk in a large, very clean saucepan and begin warming to 30°C (86°F). Meanwhile, mix the citric acid into 250 ml (9 fl oz) of the water. Stir well and add to the milk as it is warming.

» Once the milk has reached the required temperature, mix the liquid rennet into the remaining 3 tablespoons of water, stir well and add to the warm milk, stirring thoroughly for 1 minute. Cover the milk with a lid and leave for 10 minutes until a knife can make a clean cut in the curd.

» Cut the curds into a grid pattern of squares about 2 cm (¾ in), cutting vertically and then horizontally as much as possible. Stir curds for 5 minutes, very gently, while heating the curds and whey to 37°C (99°F). Continue to stir gently for 5 more minutes.

» Once the curds are clumping together, gently lift them out of the whey with a slotted spoon and place them in a ceramic or heatproof bowl, tipping out any excess liquid.

RECIPE CONTINUES ON PAGE 94

- » Now the curds need to be heated to allow them to stretch. Microwave the curds for 1 minute until the internal temperature of the curd is 57°C (135°F). If the curds are not quite hot enough, continue heating in 20 second bursts until the curds are hot enough. Once the curds are hot enough, protect your hands by using cleaned, sanitised rubber gloves. Sprinkle the curd with the salt, then fold the outer edges of the curd into the centre and keep folding and stretching until the curd is smooth and elastic.
- » Alternatively, place the curds in a colander and place in a larger bowl, then pour boiling water over the curd and allow to rest in the hot water for 1 minute. With your hands in clean and sanitised rubber gloves, remove colander from the hot water, sprinkle the salt over the curd and begin folding the outer edges of the curd into the centre. Check internal temperature of the curd. Once it reaches 57°C (135°F), you can continue folding and stretching the curd until smooth.
- » Once the curd is silky and smooth (using either method), shape into one large or two smaller balls by tearing off as much as you want and folding firmly and tucking edges under and into centre of ball (microwave for a few seconds if ball begins to harden before ball is made).
- » The buffalo mozzarella is now ready to eat. If you prefer to store it, reserve some of the whey and mix with a little salt. Immerse the cheese and chill for up to 1 week. Note, water and salt can be used if you prefer not to use the whey.

buffalo and sheep's and mixed milk cheese

SPANISH-STYLE SHEEP'S CHEESE

Makes: 4–5 medium cheeses

I love working with sheep's milk. It is rich and so lustrous as a curd—full of proteins that give the cheese its silky texture and high in fat, which gives it such a wonderful flavor.

INGREDIENTS

10 litres (10 quarts) sheep's milk
pinch of B Flora and/or MA4002
2 ml (0.006 fl oz) calcium chloride
2 ml (0.006 fl oz) vegetarian rennet
salt

METHOD

» Heat the milk to 33°C (91°F), add the cultures and stir well for 1 minute. Allow to rest for 30 minutes.

» Add the calcium chloride and stir well, then add the rennet and stir for a minimum of 1 minute and maximum of 2 minutes.

» Allow to sit undisturbed for 1 hour until the curd has set and when tested with a knife, cuts cleanly without the curd slipping off the blade (see Basics – Techniques).

» Cut the curd into 2 cm (¾ in) squares and stir briefly to ensure no large pieces remain. If you do see any large pieces of curd, quickly cut these to the approximate size of the rest of the curd. Allow the cut curds to rest for 10 minutes (this is called 'healing'). Stir well for 2 minutes, then rest for a further 5 minutes.

» Gently increase the heat of the curds and whey to 37°C (99°F) over the next 15–20 minutes, cutting the curds again until they are almost 'rice sized' and stir well intermittently so that the curds do not clump together. Allow to rest for 5 minutes so that the curds settle and the whey is covering the surface.

» Remove whey until the liquid is level with the curds, measuring how much whey you discard, then replace this volume with an equal amount of water of the same temperature (37°C/99°F). Stir gently for the next 10 minutes, maintaining the heat of 37°C (99°F).

RECIPE CONTINUES ON PAGE 97

- » Scoop the shrunken curds from the whey and divide evenly into your cheese moulds and stack them 'two high' so that the top cheese acts as a weight for the bottom cheese. After 20 minutes, swap so that the bottom cheese is now on top.
- » After 30 minutes, unstack the cheeses and turn each cheese by tipping out of the mould, turning over then returning to the mould so that the rough top surface of the cheese is now on the bottom of the mould.
- » After an hour, turn cheeses again then allow to continue to drain on a rack, overnight.
- » The next day, salt the cheese. Weigh each cheese and multiply this weight by 2%. For example, if the cheese weighs 300 g (10½ oz), you will need 6 g (⅛ oz) of salt.
- » Holding the cheese carefully, rub the salt all over the top, bottom and sides of the cheese and then return to the mould. Repeat with all other cheeses. Allow to sit on a draining rack for 24 hours, turning once during that time.
- » Prepare a large, deep container with a lid that is large enough for all the cheese to sit in without touching each other. Line with paper towel and ideally, with a plastic draining piece for the cheese to sit on.
- » Place the cheese on the rack, place the lid ajar and place in a cool environment of about 12°C (54°F). A wine fridge or cool garage or laundry is ideal. If there is no suitable place, a fridge is fine but the cheese will take longer to mature.
- » Every 3–5 days, remove the cheese and wipe out the container, replace the paper and return cheese to its box. Cover with the lid ajar and return to the cool place. Continue this process for up to 2 months until the cheese has a dry, natural rind and is firm to the touch.
- » If the cheese has developed powdery, grey mould, brush this off with a clean and dry brush then allow to sit in a well ventilated fridge for 2 days, turning after the first day.
- » Wrap the cheese in baking paper or cheese paper and place in a plastic container in a cold fridge for up to 6 months. Note, the older the cheese, the stronger the flavor.

BURRATA

Makes: 2 balls

A burrata is a stuffed mozzarella. Depending on the region in Italy, these are stuffed with either cream, mascarpone cheese, ricotta cheese or small pieces of mozzarella mixed with cream or mascarpone. Normally, Burrata would be made from scratch, with the mozzarella curd immediately stretched, filled and shaped.

In this recipe, I am presuming you have already bought or made mozzarella and now want to fill it so I've adapted the method to allow you to reheat the balls to make them malleable.

INGREDIENTS

2 fresh mozzarella balls
your choice of 2 tablespoons fresh ricotta, mascarpone, torn pieces of mozzarella or cream
fresh basil leaves, tomatoes and balsamic vinegar, to serve (optional)

METHOD

» Using 1 ball of mozzarella at a time, microwave on high for 30–45 seconds until the cheese is very hot and malleable. Quickly spread it out on a board that you have lined with plastic wrap, so that you have access to the centre of the ball.

» Place your chosen filling in the middle, then quickly stretch and fold the sides over the filling to enclose it and turn the cheese over so that the joins are on the bottom. Repeat with the second mozzarella.

» Serve with fresh basil leaves, summer tomatoes and a drizzle of aged balsamic vinegar.

BREADS

GERMAN PUMPERNICKEL BREAD

Makes: 1 loaf

Robust and full of flavor, this bread is dense enough to support a wide range of toppings such as cold meats, cheese and vegetables and is perfect as a base for interesting open sandwiches. As a bonus, it has amazingly good keeping quality and makes fantastic toast.

INGREDIENTS

450 g (1 lb) wholemeal (whole-wheat) flour
100 g (3½ oz) rye flour
115 g (4 oz) mashed potato
500 ml (17 fl oz) hot water
2 teaspoons caraway seeds
1 tablespoon salt
3 tablespoons polenta
3 tablespoons bran
2 tablespoons molasses
2 tablespoons kibbled rye
1 tablespoon yeast
egg, for glazing

METHOD

» Mix all ingredients to a sticky dough and knead well, using only as much flour as necessary to handle the dough. Since the dough is naturally sticky, avoid adding too much flour while kneading. When the dough is sufficiently smooth, shape it into a round ball and place it in an oiled bowl.

» Allow to rise in an oiled bowl for about 2 hours.

» Preheat the oven to 190°C (375°F). Lightly grease a loaf tin.

» Remove the dough from the bowl and knead briefly. Shape the dough into a round ball or place in the loaf tin.

» Allow to rise until doubled and brush with beaten egg. Bake for 40 minutes, then leave to cool on a wire rack.

PROVENÇAL FLATBREAD

Makes: 12 small breads

I had been making my own version of Fougasse before I visited a wonderful old bakery in Sarrian, in Provence. I spent the whole morning watching the baker lovingly tend his dough and then shape and cut it with great precision and flourish. The aroma as his breads baked in the 300 year old oven is something I will never be able to emulate at home but the memory is locked away forever. Begin this recipe at least one day ahead.

INGREDIENTS

Starter
300 g (10½ oz) bread flour
1 large tablespoon yeast
375 ml (13 fl oz) warm water

Dough
1½ kg (3 lb 5 oz) plain (all-purpose) flour
1½ tablespoons salt
100 ml (3½ fl oz) olive oil
1 tablespoon yeast
8 garlic cloves, minced
250 ml (9 fl oz) warm water
sea salt, to garnish

METHOD

» To make the starter, mix the flour, yeast and water together until the mixture resembles a semi-thick batter. Allow to prove covered in a non-reactive bowl for up to 3 days (8 hours minimum) to develop a lovely mature flavor.

» Mix the starter, flour, salt, yeast, garlic and half the oil with approximately 250 ml (9 fl oz) warm water to make a soft dough. Knead on a floured surface until the dough is silky smooth, adding flour as necessary until the dough is no longer sticky. Allow the dough to rise in an oiled bowl until doubled, about 2 hours.

» Divide the dough into 12 pieces and using your fingertips or a rolling pin, shape into ovals about 1 cm (½ in) thick. With a sharp knife, cut diagonal cuts through the dough and then gently stretch to open up the holes. Brush with remaining oil and sprinkle with sea salt.

» Allow to rise for 30 minutes, then bake at 225°C (425°F) for 15–20 minutes, spraying with water twice during baking (if you prefer, place a baking pan of boiling water in the bottom of the oven to create steam).

» Remove from the oven and brush once more with olive oil before they cool.

Variations: Feel free to add chopped olives, herbs, sundried tomatoes, bacon or other flavorings to the raw dough and knead through before baking for a delicious alternative.

IRISH FRUIT BREAD

Makes: 1 loaf

This traditional Irish loaf is chock-full of currants and by tradition should contain no other fruit. These days however, peel and sultanas are often substituted for some or all of the currants. Traditionally, this bread was baked over an open fire in a cast iron pot but these days is baked in a conventional oven.

INGREDIENTS

600 g (1 lb 5 oz/4 cups) unbleached bread flour
1 tablespoon yeast
2 teaspoons salt
1 teaspoon cinnamon
1 teaspoon mixed spice
90 g (3¼ oz) butter
90 g (3¼ oz) demerara sugar (or white sugar)
2 eggs
300 g (10½ oz) currants (or mixed fruit)
200 ml (7 fl oz) warm milk
50–100 ml (1¾–3½ fl oz) water (or as necessary)

Glaze
2 tablespoons sugar
2 tablespoons water

METHOD

» Place the flour in large bowl and add the salt and spices. Rub in the butter until it resembles breadcrumbs, then add the yeast, sugar, eggs, currants and milk. Mix to a thick, shaggy dough. Add as much water as necessary to achieve a sticky but manageable texture.

» Turn out and knead until the dough is no longer sticky and is very soft, smooth and elastic.

» Place in an oiled bowl and allow to rise for 2 hours, or until doubled. Turn out and shape into a high, round cob.

» Preheat the oven to 200°C (400°F). Grease a 20 cm (8 in) round cake tin.

» Place the dough in the tin and allow to rise for 45 minutes or until doubled and puffy.

» Make the glaze by boiling the sugar and water together for 2 minutes, then brush over the top of the dough.

» Bake the bread for 45 minutes, or until deep golden. Brush with more sugar syrup when removed from the oven and cool.

ZATAAR-SPICED BREADS

Makes: 2 loafs

Zataar is term given to spice mixes in many Middle Eastern countries. The varieties are as diverse as the countries themselves. This method in this recipe includes information on making Arabic Zataar spiced bread as well as Zataar spiced bread sticks, which are wonderful with cheese.

INGREDIENTS

Bread
- 375 g (13 oz) unbleached bread flour
- 375 g (13 oz) wholemeal (wholewheat) flour
- 1 tablespoon yeast
- 1 tablespoon salt
- 375 ml (13 fl oz) warm water
- 1 tablespoon olive oil

Zataar spice mix
- 1 tablespoon ground oregano
- 1 tablespoon dried oregano leaves
- 4 tablespoons dried ground thyme
- 1 tablespoon dried thyme leaves
- 1 tablespoon dried ground marjoram
- 1 teaspoon dried marjoram leaves
- ¾ cup sesame seeds
- 3 teaspoons salt
- zest of 2 lemons

METHOD

- In a large bowl, place the wholemeal flour and 300 g (10½ oz) bread flour with the yeast, salt, oil and water and mix until the dough forms a shaggy mass. If the dough is sticky, add enough of the remaining flour until the dough is kneadable.
- Knead for 10 minutes or until silky smooth and elastic. Allow the dough to rise in an oiled bowl until doubled.
- Meanwhile, mix all zataar ingredients until thoroughly combined.
- Turn the dough out from the bowl and knead briefly.
- To make Arabic breads, divide the dough into two and roll out to a rectangle or oval about 1 cm (½ in) thick. Brush with olive oil and sprinkle generously with spice mix. Allow the breads to rise for 30 minutes, then score deeply in a diagonal pattern.
- Bake at 210°C (415°F) for 20–25 minutes, or until golden. (To make the bread more moist, spray with water during baking).
- To make the spiced breadsticks, roll the dough out to a rectangle about 5 mm (¼ in) thick. Brush with olive oil and sprinkle with spice mix.
- Using a pizza wheel or knife, cut the dough into long strips. Pick up both ends and twist to form a long rope. Place on a baking tray and continue until all the dough is shaped in this fashion.
- Bake at 230°C (450°F) for 15 minutes, or until crisp and golden.

ROSEMARY & SULTANA PANINI

Makes: 16 small paninis

Aromatic and bursting with flavor, these crusty rolls complement anything they are served with. The unusual combination of sultanas and rosemary creates a perfect marriage of flavors and whenever I serve these delicious little rolls, they are always very popular. Perfect on a versatile cheese board, especially with blue cheese!

INGREDIENTS

500 g (1 lb 2 oz) unbleached bread flour
1 tablespoon sugar
1 tablespoon freeze-dried yeast
1 teaspoon salt
350 ml (12 fl oz) warm milk
   (or water if preferred)
3 tablespoons olive oil
4 tablespoons rosemary leaves
4 tablespoon sultanas
1 tablespoon olive oil

METHOD

- Place the flour, yeast, salt and sugar in a large mixing bowl and add the warm milk and olive oil. Stir the mixture well, then turn out onto a floured surface and knead for about 8 minutes or until the mixture is smooth and elastic.
- Allow the dough to rise in an oiled bowl for about 1 hour, or until doubled in size.
- Meanwhile, heat the extra olive oil and cook the rosemary and sultanas for 1–2 minutes or until the mixture is fragrant. Allow the mixture to cool.
- Turn the dough out onto a floured surface and knead briefly, flattening out the dough with your fingertips. Spread the rosemary mixture over the flattened dough, then roll up the dough tightly to enclose the rosemary filling. Knead the dough thoroughly to incorporate the rosemary and sultanas, then cover the dough for 5 minutes to allow it to relax.
- Divide the dough into 16 even pieces, then shape the dough into balls the size of a golf ball. Place them on an oiled baking tray. Cover and leave to rise for about 30 minutes or until doubled. Preheat the oven to 200°C (400°F).
- Dust the panini with flour, then bake for 18 minutes, or until golden brown. Remove the breads from the oven and allow to cool on a wire rack.

HERB FOCACCIA

Makes: 1 flatbread

Delicious, easy to make and so versatile, this is a very good basic focaccia recipe that can be adapted to include olives or sundried tomatoes if you prefer. It freezes very well so make two from this batch and pop one in the freezer for another day!

INGREDIENTS

600 g (1 lb 5 oz/4 cups) unbleached bread flour
1 tablespoon yeast
1 tablespoon salt
4 tablespoons oil
2 garlic cloves, minced
2 teaspoons chopped rosemary
1 small handful chopped flat-leaf (Italian) parsley
20 basil leaves, chopped
2 tablespoons chopped dill
400 ml (14 fl oz) water
2 tablespoons oil, extra
sea salt and cracked pepper

METHOD

» Mix the flour, yeast and salt together with all the fresh herbs. Mix the water, oil and garlic together then add to the flour mixture and mix the ingredients together to form a firm dough. Knead until the dough is smooth and elastic, about 10 minutes.

» Place the dough in an oiled bowl, covered with plastic wrap, to rise until doubled.

» When the dough has risen, remove the dough from the bowl and press onto a greased oven tray to a thickness of about 1 cm (½ in).

» Drizzle with the extra oil and sprinkle with sea salt and cracked pepper if desired, then dimple deeply with your fingertips so that the oil fills the dimples.

» Allow to rise for 30 minutes, then bake at 200C (400°F) for 30 minutes or until golden brown. Remove from the oven and cool on a wire rack.

Note: If you prefer the herbs on top of the bread, reserve them until after you shape the dough. Press the dough onto the baking tray then sprinkle the mixed herbs over the surface. Drizzle with olive oil, salt and cracked pepper and bake as above.

RAISIN RYE STICKS WITH WALNUTS

Makes: 4 sticks

This is without doubt one of my favorite breads. I adore the flavor of rye flour with the crunch of nuts and the sweetness of raisins. The glaze of molasses adds richness and color to an already scrumptious loaf. It is best sliced thinly and served with a cheese and fruit platter, particularly with a mild blue cheese and firm pears. Of course, it is so irresistible straight out of the oven, that it might not even last long enough to accompany the cheese.

INGREDIENTS

Starter
200 g (7 oz) rye flour
500 ml (17 fl oz) warm water
2 tablespoons yeast

Dough
300 g (10½ oz) unbleached bread flour
200 g (7 oz) rye flour
2 teaspoons salt
1 teaspoon yeast
1½ cups rye starter
350–400 ml (12–14 fl oz) warm water
200 g (7 oz) walnut pieces
200 g (7 oz) raisins
2 tablespoons treacle
2 tablespoons molasses
2 teaspoons caraway seeds

METHOD

» First, make the starter. Mix the rye flour, water and yeast together with a wooden spoon and allow to ferment overnight in a covered bowl.

» The next day, make the bread. Place the flours, salt, yeast, water, treacle and rye starter in a large bowl and begin to knead, adding a little extra flour if the dough seems to wet to handle (the dough is traditionally a little sticky).

» Transfer the dough to an oiled bowl for 1 hour to rise.

» Meanwhile, mix the molasses with 2 tablespoons of hot water.

» Divide the dough into six equal pieces and shape these into long, skinny 'ropes'. Transfer the ropes to a baking tray and brush generously with the molasses glaze. Leave to rise for about 1 hour or until doubled in size.

» Preheat the oven to 200°C (400°F). Brush the 'ropes' with any remaining glaze and slash the surface. Spray with water and sprinkle with caraway seeds, if desired.

» Bake for 30 minutes, or until cooked through and crisp.

FRENCH SOURDOUGH WITH CARAMELIZED ONIONS

Makes: 2 loafs

The inspiration for this recipe comes from a similar bread bought in the food markets of St. Remy de Provence. It has a delicious, sweet-sour taste that marries beautifully with the sweetness of the caramelized onions. This is a fantastic choice when serving a Ploughman's Platter (a selection of cheese, cold cuts, relish and bread).

INGREDIENTS

Starter
- 250 g (9 oz) plain yoghurt
- 150 g (5½ oz) wholemeal flour
- 3 tablespoons warm water
- 1 teaspoon sugar

Dough
- 3 tablespoons warm water
- 1 teaspoon sugar
- 300–450 g (10½–1 lb) wholemeal flour
- 1 tablespoon dried yeast
- 1½ teaspoons salt
- 1 teaspoon baking soda
- 4 large onions, sliced
- 50 g (1¾ oz) butter

METHOD

One day before:
» In a large mixing bowl, combine 150 g (5½ oz) wholemeal flour, the starter, 1 teaspoon sugar and 3 tablespoons warm water. Mix well and set aside to ferment for 24 hours.

The next day:
» Melt the butter in a large saucepan and add the sliced onions. Stir to coat with the butter and cook over medium heat until the onions are translucent. Cover the saucepan with a lid and continue to cook on a low heat for about 40 minutes, or until the onions are golden brown. Set aside to cool.

» Mix the yeast, water and sugar together and allow to sit for 5 minutes. Mix this mixture into the prepared starter dough, with the salt and the caramelized onions. Slowly add more flour until the dough forms a shaggy mass. Tip the dough out onto a floured board or bench. Begin kneading the dough adding as much flour as necessary. When the dough is quite smooth and manageable, allow the dough to rise for 30 minutes.

» Remove the dough from the bowl and divide in half. Shape each piece of dough into a flat oval loaf, about 1½ cm (⅝ in) thick, using your fingertips to add texture to the dough.

» Drizzle the dough with some olive oil and allow to rest again for 30 minutes. Spray the dough with water and bake in a preheated oven at 200°C (400°F) for 25–30 minutes, or until crusty and golden.

Note: Alternatively, if you prefer a more rustic version, reserve the onions until the bread is shaped then arrange over the top of the bread, covering the dough. Rise and bake as above.

MIDDLE EASTERN PITA BREAD

Makes: 16 pita

Pita are very rewarding to make. They are quick, easy and delicious. If you find they do not puff up the first time, don't worry, they will still be delicious. This dough can be stored in the fridge for up to a week so you can have fresh pita whenever you are ready.

INGREDIENTS

375 g (13 oz) wholemeal (whole-wheat) flour
375 g (13 oz) unbleached bread flour
1 tablespoon yeast
1 tablespoon salt
1 tablespoon olive oil
625 ml (2½ fl oz) warm water

METHOD

- In a large bowl, place the wholemeal flour and 300 g (10½ oz) bread flour with the yeast, salt, oil and water and mix until the dough forms a shaggy mass. If the dough is sticky, add enough of the remaining flour until the dough is kneadable.
- Knead for about 10 minutes, or until silky smooth and elastic. Allow the dough to rise in an oiled bowl for about 1 hour, or until doubled.
- Turn the dough out and divide into 16 equal pieces. Cover the dough pieces with a towel and work on one or two at a time. Roll each piece out to a circle about 20 cm (8 in) in diameter and about 5 mm (¼ in) thick, taking care that there are no creases in the dough.
- Carefully place 2 circles of dough on a tray (do not overlap) and repeat with a second tray. Bake both trays together at 230°C (450°F) for 3–4 minutes, or until the breads 'balloon'. Do not allow to burn or scorch. When removing the breads from the oven, stack them on top of each other to cool.
- Bake all the breads in the same fashion and wrap them in a clean cloth or tea towel to keep them moist and tender until serving.

ITALIAN GRISSINI

Makes: 16–20 medium sticks

Grissini are probably the easiest of the Italian breads to make. They require very little rising or shaping and are very quick to bake. Traditionally, their shape should be fairly rough and knobbly—not perfectly straight like those available in Italian bistros these days. The shape is really irrelevant, the taste of these breadsticks is rustic and delicious, and perfect for dipping and spreading with oozy cheeses.

INGREDIENTS

500 g (1 lb 2 oz) unbleached bread flour
1 tablespoon yeast
1 tablespoon malt syrup (or golden syrup)
2 teaspoons salt
310 ml (10¾ fl oz) warm water
2 tablespoons olive oil
sesame seeds, poppy seeds and semolina flour, for sprinkling

METHOD

» Stir the malt syrup and oil into the warm water and then add to the flour together with the yeast and salt. Mix until the dough comes together and then turn out onto a floured bench and knead until the dough is smooth and elastic.

» Shape the dough into a ball and brush the surface with oil lightly, and allow the dough to rise for about 1 hour.

» When ready to shape, deflate the dough and flatten. Cut the dough into four equal size pieces and then cut each piece into strips about 1 cm (¾ in) thick. Simply shape these by holding each end of the strip and pulling gently so that the strip is about 20 cm (8 in) long. Alternatively, you can shape these more 'perfectly' by dividing the dough into 20 small pieces and rolling each of these out to form a long rope about 30 cm (12 in) long.

» Spray the ropes with water, then roll in sesame seeds or poppy seeds. Place in a grissini tin or on a baking tray. There is no need to let the grissini rise. Spray with water and bake at 220°C (425°F) for 20 minutes or until golden and crisp, then transfer to a wire rack to cool.

COOKING WITH CHEESE

BELGIAN MUSSELS WITH BLUE CHEESE

Serves: 4 as an entrée

Fresh briny mussels, lightly steamed and served with crumbled blue cheese are a marriage made in heaven. Choose a blue cheese that is not too pungent.

INGREDIENTS

2 kg (approx. 5 lb) fresh blue mussels
2 tablespoons olive oil
1 brown onion, peeled and diced
4 shallots, peeled and diced
2 bacon slices, diced (optional)
250 ml (9 fl oz) white wine
juice of 1 lemon
1 large handful baby English spinach, washed
150 g (5½ oz) blue cheese, crumbled
sea salt, to taste
freshly ground pepper, to taste
finely sliced spring onions (scallions), to serve
chopped flat-leaf (Italian) parsley, to serve

METHOD

» Heat the oil in a large deep saucepan over medium heat. Add the onion, shallots and bacon (if using) and cook until golden.

» Add the mussels and white wine, stir quickly, then cover and cook for 5 minutes or until the mussels are all opened. Set aside any mussels that haven't opened.

» Add the blue cheese and stir into the broth to melt. Add the spinach and cover for 1 minute.

» Season to taste and serve with extra crumbled blue cheese, thinly sliced spring onions and chopped flat-leaf parsley.

Note: It is commonly thought that any unopened mussels should be discarded because they are dead, but this is usually not correct. It is simply because they didn't receive enough heat in the pan. Return any unopened mussels to remaining broth (or a little extra stock) and cover. Bring to the boil and allow to cook for 2–3 minutes. They should all open. Any that continue to remain shut should be discarded.

SAGANAKI

Makes: 1 slice per person

Mediterranean Saganaki is simply an appetiser of pan-fried or grilled Haloumi, served with a squeeze of lemon and some salad leaves, if you like. You can dress it up with olives, a salad of cucumber and tomato and oregano or you can just enjoy it plain…either way, it is delicious! We haven't specified amounts in the ingredients as you just cook as much or as little as you wish.

INGREDIENTS

haloumi, cut into slices 5 mm (⅕ in) thick
olive oil
lemon wedges, to serve

METHOD

- » Heat a non-stick frying pan over high heat until almost smoking, then add a drizzle of olive oil for each piece of cheese to be cooked.
- » Add the cheese and cook until you can see golden color around the edge of the cheese.
- » Remove the pan from the heat and use a spatula to gently turn the cheese, taking care not to disturb the crust.
- » Cook the other side until golden, then remove from the pan and allow to sit for 1 minute until the sizzling subsides, then remove to a plate.
- » Serve over salad with a squeeze of lemon.

FRENCH-STYLE GOAT CHEESE SALAD

Makes: 1–2 slices of baguette per person

Standard lunch fare throughout France, a salad with grilled goat cheese croutons is one of my favorite meals. You can substitute the asparagus for other vegetables such as artichokes, roasted bell capsicum (peppers), blanched green beans or other vegetables you may prefer to include. The measurements aren't exact so you can make as much or as little as you wish.

INGREDIENTS

assorted fresh salad greens
cherry tomatoes, halved
fresh asparagus, blanched
1–2 slices of French baguette, per person
ashed goat cheese or goat cheese of your choice
toasted walnuts, to serve

Dressing

2 tablespoons olive oil
1 tablespoon balsamic vinegar
1 tablespoon mayonnaise
1 tablespoon red wine vinegar
salt and pepper, to taste

METHOD

» Arrange the salad leaves on a serving plate. Add the cherry tomatoes and arrange the asparagus spears.
» Toast the baguette slices. Top each slice with a diagonal slice of the ashed goat log or if the cheese is spreadable, spread gently on the toasted bread then grill (broil) until the cheese starts to bubble and color slightly.
» Whisk the dressing ingredients and season with salt and pepper, then drizzle over the salad. Top with 1–2 slices of the grilled goat cheese croutons and serve immediately, sprinkled with walnuts.

Double-Baked Cheese Soufflé

Serves: 4

I always think it's a shame that people are scared of baking soufflés. They are such a delight to make and so rewarding. This one is easy too because you make it before you need it, then simply re-bake to serve. As if by magic, it puffs up again just before serving and is very impressive!

INGREDIENTS

60 g (2¼ oz) butter, plus extra for soufflé dishes
2 tablespoons breadcrumbs
60 g (2¼ oz) plain (all-purpose) flour
350 ml (12 fl oz) milk, heated
150 g (5½ oz) grated gruyère, cheddar, blue cheese or other firm cow's milk cheese
2 tablespoons chopped chives
sea salt and freshly ground black pepper
4 eggs, separated
grated parmesan, to serve
mild paprika, for sprinkling
green salad, to serve (optional)
tomato chutney, to serve (optional)

METHOD

» Preheat the oven to 180°C (350°F). Lightly butter four 250 ml (9 fl oz/1 cup) soufflé moulds. Dust with the breadcrumbs and set aside.

» Melt the butter in a saucepan, add the flour and stir for 2 minutes over medium heat until the mixture has a nutty aroma and is golden brown. Do not allow to burn.

» Add the heated milk gradually, stirring constantly. Add the grated cheese, chives, salt and pepper, and cook gently for 5 minutes, stirring until the mixture is thick and fragrant.

» Cool for 10 minutes, then beat in the yolks, one at a time.

» Whip the whites with a pinch of salt until soft peaks form, then fold in gently with the cheese mixture.

» Fill each mould to three-quarters full, place in a baking dish and half-fill the dish with hot water. Bake for 20 minutes, or until golden and risen. Remove from the oven and set aside until needed or turn out onto a baking tray to cool (the soufflé will deflate, do not worry about this). You can re-bake the soufflé in moulds or on a baking tray if you prefer to serve on a salad.

» To serve, sprinkle each soufflé with some grated parmesan cheese and sprinkle with paprika. Bake at 220°C (425°F) for 6 minutes, or until puffed and golden. Serve with a green salad and a tomato chutney.

GOAT CHEESE MUSTARD GOUGÈRES

Makes: 30–40 small Gougères

These lovely little choux puff pastries are so easy to make, despite the belief that choux pastry is tricky. These can be flavored with any cheese but I love the acid flavor of lactic goat cheese in the rich dough. A little nutmeg adds another element.

INGREDIENTS

250 ml (9 fl oz) water
125 ml (4 fl oz) milk
1–2 teaspoons dijon mustard
125 g (4½ oz) unsalted butter
150 g (5½ oz/1 cup) plain (all-purpose) flour, plus an extra 2 tablespoons
4 large eggs
sprinkling of nutmeg
130 g (4¾ oz) crumbled goat's cheese of your choice
salt, to taste
1 tablespoon sesame and poppy seeds

METHOD

» Preheat the oven to 210°C (415°F).
» In a saucepan, melt the butter with the milk, water and mustard until simmering.
» Bring to the boil and add the flour, constantly stirring for 2 minutes or until well incorporated, and the mixture leaves the side of the pan as you beat.
» Remove the pan from the heat. Transfer the mixture to a mixing bowl and cool slightly.
» Place the dough in an electric stand mixer with a paddle attachment. Set the speed to medium, and add the eggs, one at a time, beating well after each addition. Once the mixture is thick and glossy, add the goat cheese and nutmeg.
» Once the dough is mixed, scoop or pipe mounds of dough (1 tablespoon each) onto a baking tray, spaced an equal distance apart. Sprinkle with sesame, poppy or both and bake for 8 minutes. Rotate the pan and bake for 5–7 minutes, or until puffed and golden brown. Serve warm.

Note: To make larger pastries or different shapes, simply allow extra time for cooking and bake until puffed and golden.

FRENCH ONION SOUP

Serves: 4–6

Such a classic dish—well-made French onion soup is substantial enough to be a whole meal and should be topped with a crouton, brimming with melted cheese. Serve the soup in ovenproof bowls so that the cheese can be grilled in the soup, just before serving.

INGREDIENTS

50 g (1¾ oz) butter
2 tablespoons olive oil
6 large onions, sliced
2 garlic cloves, crushed
4 tablespoons red wine
1 tablespoon plain (all-purpose) flour (optional)
1 litre (35 fl oz) beef stock or consommé
4 tablespoons dry sherry or port
1 French-style baguette
200 g (7 oz) Comté cheese, grated
100 g (3½ oz) gruyère, grated
flat-leaf (Italian) parsley, chopped, to garnish

METHOD

» Heat the butter and olive oil in a large heavy-based saucepan. Add the sliced onions, then the garlic. Sauté over high heat until the onions begin to color around the edges, then reduce the heat and cook gently for 15 minutes or until the onions are golden.

» Increase the heat then 'deglaze' the pan by adding the red wine. Cook over high heat, stirring until the alcohol evaporates and the liquid disappears. If you are using the flour, add now and stir in thoroughly.

» Add the beef stock or consommé and sherry or port and bring to the boil, stirring gently. Simmer, covered, for about 45 minutes or until the soup is rich and deep brown in color. Season to taste with salt and pepper and add a little water if the soup is too rich. Add the parsley and stir through.

» To make the croutons, grill (broil) the slices of bread until golden on both sides. Mix the two cheeses together, then pile generously on each piece of toast and return to the grill (broiler) and cook until cheese is well melted and golden.

» To serve, ladle the piping hot soup into each bowl. Slip a grilled crouton on the surface, then return the bowls to the grill (broiler) for a further 1–2 minutes or until the cheese is bubbling. Serve immediately.

GOAT CHEESE CROSTADA

Serves: 4–6

This freeform tart features rustic pastry encasing a delicious Mediterranean-style filling of roasted vegetables.

INGREDIENTS

Filling
2 tablespoons olive oil
4 large red onions
2 red capsicums (bell peppers)
2 rosemary sprigs
2 tablespoons brown sugar
1 tablespoon balsamic vinegar
250 g (9 oz) goat cheese of your choice

Pastry
300 g (10½ oz/2 cups) plain (all-purpose) flour
2 teaspoons rosemary
2 tablespoons polenta grains
1 teaspoon baking powder
240 g (8½ oz) cold butter, diced
2 tablespoons cold water

METHOD

» To make the filling, slice the onions into thick rounds or wedges. Slice the capsicum flesh into wedges and discard the seed core.

» Toss the onion and capsicum pieces with the oil, rosemary, brown sugar and balsamic vinegar and bake at 190°C (360°F) for 50 minutes, tossing during the cooking period. When the vegetables are golden and soft, remove from the oven and allow to cool.

» In a food processor, 'pulse' the flour, salt, rosemary, polenta and baking powder to mix. With the motor running, add the butter cubes and process briefly. Add just enough water to help the mixture hold together to reach the ball stage. Remove from the processor and flatten to a disk. Chill until required.

» To assemble, roll out the dough to 4 mm (½ in) thick, round shape. Pile the filling in the centre of the dough then gently spread to within 6 cm (2½ in) of the edge. Sprinkle the goat cheese over the filling. Lift up the edge and fold forward to encase some of the filling and pinch to create a freeform crust.

» Place on a non-stick baking tray and bake at 220°C (425°F) for 10 minutes, then reduce the heat to 180°C (350°F). Bake for 45 minutes, or until pastry is crisp and golden.

» Serve hot, warm or cold in wedges with a salad.

HALOUMI POTATO PIE

Makes: 1 large or 4 small

Hearty and delicious, this rich potato pie makes good use of haloumi in the layers of potato, and the grated cheese on top creates a lovely crusty finish!

INGREDIENTS

50 g (1¾ oz) butter, softened
2 kg (4 lb 8 oz) washed pink potatoes, skin on, thickly sliced
4 brown onions, peeled and thinly sliced
200 g (7 oz) haloumi
250 g (9 oz) thick cream (or evaporated milk)
salt and pepper, to taste,
ground paprika, to sprinkle

METHOD

» Grease a casserole dish by rubbing softened butter all over the base and sides. Preheat the oven to 200°C (400°F).

» Create a layer of potato slices, onion, salt and pepper and a little grated haloumi, then another layer, ending with any remaining potato slices.

» Dilute the cream or evaporated milk with 100 ml (3½ fl oz) warm water then drizzle over. Grate or thinly slice remaining haloumi over the top, then sprinkle with paprika.

» Bake at for 45 minutes, or until golden and bubbling. Serve hot.

Note: For individual pies, follow instructions but use 4 small pie dishes and bake for 30–35 minutes.

RISOTTO OF FOUR CHEESES

Serves: 4–6 as entrée

This sublime dish is my interpretation of a classic Italian pasta sauce, usually served with fettuccine. A risotto made with four cheeses is rich so only serve in small portions with a mixed green salad to cut the richness.

INGREDIENTS

1 tablespoon butter
1 tablespoon olive oil
1 large onion, finely chopped
400 g (14 oz) arborio rice
250 ml (9 fl oz) white wine
1 litre (35 fl oz) vegetable stock
90 g (3¼ oz) fresh goat curd
90 g (3¼ oz) mild blue cheese
90 g (3¼ oz) mild cow's milk cheese, grated
60 g (2¼ oz) hard goat cheese or Parmesan
fresh parsley and chives, to garnish
freshly ground black pepper, to season

METHOD

» Heat the butter and olive oil in a saucepan and add the onion. Sauté gently until the onions are softened and but not colored. Add the rice and stir to coat. Add the wine and stir well, allowing the liquid to be absorbed.

» Begin adding the stock, a ladle at a time and stir, allowing each addition to be absorbed before the next addition.

» When adding the last quantity of stock, add all the cheeses and stir to combine. Cover and remove the risotto from the heat. Allow to sit for a moment or two so that the cheeses can melt.

» Fold through plenty of fresh parsley and ground black pepper to taste. Serve immediately.

ROASTED BEETROOT SALAD WITH PERSIAN FETA

Serves: 6–8 as a side dish

This gorgeous summer salad features earthy beetroot and spinach, the freshness of marinated goat cheese and the crunch of nuts to create a perfectly light lunch.

INGREDIENTS

1 kg (2 lb 4 oz) fresh beetroot, trimmed and scrubbed
2 tablespoons olive oil
3 small Spanish onions, cut into thin wedges
1 small red onion, finely chopped
180 g (6½ oz) green beans, trimmed
1 handful baby English spinach leaves
120 g (4¼ oz) toasted pine nuts or walnuts
120 g (4¼ oz) marinated goat feta

Dressing

2 tablespoons olive oil
1 tablespoon balsamic vinegar
1 tablespoon red wine vinegar
salt and pepper, to taste

METHOD

» Boil the beetroot in salted water for 30 minutes until barely tender, then drain and quarter. Place in a baking dish with the wedges of Spanish onion with the olive oil and salt and pepper to taste. Bake at 220°C (425°F) for 30 minutes or until cooked through.

» Remove from the oven and add the green beans then return to oven for 5 minutes.

» Remove from the oven and add the chopped onion and green beans and return to the oven for 5 minutes.

» Toss with spinach leaves and the dressing, then add the marinated feta and nuts sprinkled over the top to serve.

» Serve warm or cold.

RISOTTO WITH SPINACH & BLUE CHEESE

Serves 4–6

This risotto is quite rich and would be perfect as an entrée or a light lunch, served with a green salad.

INGREDIENTS

2 tablespoons olive oil
2 garlic cloves, minced or 2 teaspoons minced garlic
1 onion, finely chopped
400 g (14 oz) arborio rice
125 ml (4 fl oz) white wine
300 g (10½ oz) spinach, cooked and chopped or 1 pack frozen spinach
2 handfuls washed baby spinach leaves
1 litre (35 fl oz) vegetable stock
210 g (7½ oz) soft blue cheese of your choice, preferably a soft goat blue
freshly ground pepper, to taste
fresh parsley or chives, to garnish

METHOD

» Bring the stock to simmer and set aside.

» Sauté the onion and garlic in the olive oil over medium heat and cook briefly until softened. Do not allow to turn golden. Add the rice and stir to coat, ensuring all the grains are glossy. Add the wine and stir thoroughly, until all the wine has been absorbed—about 2 minutes.

» Add the chopped fresh or frozen spinach and one ladle of stock and stir well, ensuring that the spinach is well distributed.

» Continue cooking over medium heat, adding stock ladle by ladle and stirring between each addition. When adding the last ladle of stock, add the baby spinach leaves and the gorgonzola, cut into small pieces, and stir to combine. Cover and immediately remove from the heat.

» Allow to sit for 3–4 minutes, then remove the lid. Stir once more to incorporate wilted spinach, cheese and remaining liquid. Serve immediately, garnished with some chopped fresh parsley or chives and plenty of freshly ground black pepper.

SWEET POTATO & PEANUT SALAD WITH HALOUMI

Serves: 8 as a side dish

Full of texture and flavor, I love to serve this salad as part of buffet lunch or dinner. It is also perfect as an accompaniment to a barbecue.

INGREDIENTS

2 kg (4 lb 8 oz) sweet potato, peeled
20 garlic cloves
1 red onion, minced
1–2 small red chillies, minced
1 large handful fresh herbs of your choice e.g. coriander, parsley, dill, chives or a mixture
300 g (10 oz) roasted peanuts
200 g (7 oz) haloumi
6 tablespoons olive oil
2 tablespoons balsamic vinegar
salt and freshly ground pepper, to taste

METHOD

» Preheat the oven to 220°C (425°F).
» Peel and cut the sweet potato into large chunks. Toss with 2 tablespoons olive oil, salt and pepper to taste.
» Place the sweet potato in a large baking dish and bake for about 45 minutes. Add the garlic cloves and stir, then continue cooking for 30 minutes or until the sweet potato is tender and golden around the edges. Remove from the oven and keep warm.
» Mix the red onion and minced chilli with the fresh herbs and combine with the sweet potato.
» Meanwhile, heat a little olive oil in a non-stick frying pan and cook the haloumi until deep golden brown. Turn and cook the other side, then cut into strips. Set aside.
» Whisk the remaining 4 tablespoons olive oil with the balsamic vinegar and toss with the pumpkin mixture. Add the peanuts and haloumi strips. Toss once more and serve.

RICOTTA-STUFFED EGGPLANT ROLLS

Serves: 6

This elegant entrée is bursting with flavor and just beautiful to look at. You can easily prepare these rolls 8 hours ahead, then chill until ready to serve which makes them perfect for a do-ahead dinner of buffet. They can be as large or small as you wish, depending on the size of your eggplant (aubergine).

INGREDIENTS

2 large, smooth eggplant (aubergine)
1–2 tablespoons olive oil
4 large, red (or yellow) capsicums (bell peppers)
210 g (7½ oz) Persian feta, drained
210 g (7½ oz) firm, fresh ricotta cheese
20 basil leaves
10 mint leaves
1 bunch of rocket or handful of baby English spinach
salt and pepper, to taste
fresh pesto, to serve

METHOD

- Slice each eggplant vertically into six even slices. Salt lightly and allow the bitter juices to drain for 30 minutes. Rinse briefly and pat dry.
- Slice sides off each capsicum so that you have 12 large pieces. Place the capsicum slices under a hot grill (broiler) and cook until the skins are blackened. Place the blackened capsicum into a plastic bag and seal, allowing them to steam for 30 minutes. When cool enough to handle, open the bag and gently peel off and discard the black capsicum skins, allow the capsicum flesh to cool.
- Lightly brush or spray the eggplant slices with oil then chargrill in a grill pan or barbecue until the eggplant pieces are golden, then turn and brush or spray with olive oil. Grill the other side until golden, about 3 minutes each side. Allow to cool.
- Meanwhile, in a bowl or food processor, combine the feta cheese, ricotta cheese and herbs with salt and pepper to taste. (If using a processor, take care not to over-process, the cheese mixture should be spreadable but not smooth.)
- To assemble, lay each eggplant slice on a flat surface and top with a large piece of capsicum. Place a generous spoonful of the cheese filling on top of the capsicum, near the bottom end closest to you. Add some herb and rocket leaves over the cheese, then roll up firmly but gently, making sure that some of the herbs protrude from each end. Serve drizzled with parsley pesto.

TOMATO & MOZZARELLA FRITTATA

Serves: 4–6

A frittata is a kind of thick omelette, filled with chunky ingredients. It is usually served at room temperature cut into fingers or slices, often as part of an antipasto platter. This one is particularly good when tomatoes and herbs are full of summer flavor.

INGREDIENTS

- 1 tablespoon butter or olive oil
- 10 large eggs
- 5 roma (plum) tomatoes
- 8 sundried tomatoes
- 1 tablespoon tomato paste (concentrated purée)
- 1 large carrot, grated
- ½ bunch of chives
- ½ cup coriander (cilantro) leaves
- 2 tablespoons chopped sage
- 2 tablespoons flat-leaf (Italian) parsley
- 2 buffalo mozzarella balls, sliced
- 2 tablespoons fresh basil pesto
- salt and pepper, to taste

METHOD

» Finely dice the tomatoes and sprinkle with 1 teaspoon of salt. Allow to drain in a sieve. Crack the eggs into a large bowl and add the tomato paste, whisking until it has dissolved. Add the chopped sundried tomatoes, grated carrot, chives, coriander, sage and the drained tomatoes and mix well.

» Heat an ovenproof pan or baking dish, add the oil, then add the egg mixture, stirring in the pan until the eggs begin to set. Cover with a lid and cook over medium heat for 5 minutes, until the eggs have cooked underneath.

» Arrange the sliced mozzarella over the surface, then transfer the ovenproof pan or frying pan to the oven and cook at 180°C (350°F) for 30 minutes, or until the frittata is cooked and a little puffed on the surface.

» Serve warm or cold with a tomato chutney or relish if desired.

FILO CHEESE TRIANGLES

Makes: 12–16 pastries

Golden, crunchy parcels of cheese and herbs are irresistible! I love these with their mixture of Persian feta and ricotta and spinach!

INGREDIENTS

- 180 g (6 oz) feta, drained
- 210 g (7 oz) fresh ricotta cheese
- 200 g (7 oz/1 cup) cooked spinach, chopped and pressed to remove liquid
- ½ cup chopped flat-leaf (Italian) parsley
- ¼ teaspoon ground nutmeg
- 375 g (13 oz) packet filo pastry sheets, defrosted
- 90 g (3¼ oz) unsalted butter, melted
- salt and pepper, to taste
- sesame seeds, for sprinkling

METHOD

» Preheat the oven to 220°C (425°F). Lightly butter a baking tray.

» In a bowl, mix the crumbled feta cheese, ricotta, spinach, parsley, nutmeg, salt and freshly ground pepper, to taste.

» Place one sheet of filo pastry on a flat surface, brush sparingly with melted butter and cover with another sheet of pastry. Brush with a little more butter and cover with a third sheet of pastry. Using a sharp knife, cut the pastry into four even, long strips.

» Place a large spoonful of cheese mixture at one end of the pastry strip and fold over the corner to form a triangle. Continue folding the pastry over to complete and reinforce the triangle, enclosing all the filling.

» Place seam side down on the baking tray. Repeat with all remaining pastry and cheese filling.

» Brush the pastries with melted butter and sprinkle with sesame seeds. Bake for 12–15 minutes, or until golden brown and puffed.

Note: If easier, you can shape these into cigar shapes instead of triangles. To do this, butter and layer pastry as above but cut into 3 long strips instead of 4. Place the filling along the short edge of the strip and shape with your fingers to be log shaped then simply roll the filling up in the pastry, using your hands to make an even log (cigar) shape. Bake as above.

WITLOF SALAD WITH APPLES, BLUE CHEESE & PECANS

Serves 6

This is a modern version of the classic Waldorf Salad, created in 1893 for a charity event at the famed Waldorf Astoria Hotel in New York. For lovers of blue cheese, this salad is close to perfection. For those who don't find blue cheese appealing, use fresh goat's cheese. Serve this salad on the day it is made for best results

INGREDIENTS
5 witlof (Belgian endive) heads
200 g (7 oz) young rocket (arugula) leaves
1 red delicious apple, cored, quartered, thinly sliced
1 granny smith apple
120 g (4¼ oz) coarsely chopped pecans, toasted
100 g (3½ oz) crumbled blue cheese, such as Gorgonzola or Stilton
3 tablespoons olive oil
3 tablespoons walnut oil
3 tablespoons sherry wine vinegar
1 large shallot, minced

METHOD
- Cut the witlof in half, lengthwise, then lay the witlof cut side down on a board and cut the leaves into thin strips.
- Thinly slice the unpeeled apples and toss with the lemon juice.
- Combine witlof strips, apple slices, rocket, toasted pecans and blue cheese in a large bowl.
- Whisk the oils, vinegar and shallot in small bowl, then season to taste with salt and pepper.
- Drizzle the dressing over the salad and toss thoroughly. Serve immediately.

PROVENÇAL TOMATO & GOAT CHEESE TARTS

Makes: 16 small tarts

The delicious combination of tomato, goat's cheese and olives—all encased in a crisp and buttery pastry shell—is a winner no matter what the occasion!

INGREDIENTS

4 sheets puff pastry
250 g (9 oz) tomato-based pasta sauce
250 g (9 oz) cherry tomatoes, chopped or sliced
75 g (2½ oz) chopped black olives (optional)
120 g (4 oz) goat's cheese (I like to use marinated Persian feta)
basil or flat-leaf (Italian) parsley, to serve

METHOD

- Defrost the pastry.
- Preheat the oven to 210°C (415°F). Line a baking tray with baking paper.
- Divide each sheet of pastry into four even squares. Fold each edge over to form a rim on each pastry square, pressing down on overlapping corners. Using a fork, prick the base of the pastry squares several times.
- Transfer the pastry squares to the lined baking tray so that they are not touching each other.
- Spoon some tomato sauce into the centre of each square and top with chopped tomato pieces, some goat cheese, olives and fresh herbs.
- Bake for 15–20 minutes or until pastry is golden and filling is bubbling a little.
- Add some more fresh herbs and serve hot, warm or cold.

GLOSSARY

Bacterial brine – A brine containing brevi linens, a culture used in 'washed rind' cheese to help develop the typical orange/red sticky rind.

B Flora – One of the most common mesophilic cultures used in cheese making.

Blue mould spores – Blue mould, usually derived from French Roquefort, used to create a blue rind and veins in cheese.

Calcium chloride – Calcium chloride is a clear liquid, used in cheesemaking to assist with proper curd formation and to restore calcium levels in milk to their original level prior to pasteurisation.

Charcoal ash – A black powder made from vegetable matter, traditionally used to protect curd from insects but also assists in creating a balanced surface for mould growth in cheese.

Flora Danica – The commercial name for B Flora cultures.

Geotrichum 13 (GEO) – A secondary culture used to develop yeast or mould like growth on the surface of cheese.

Geotrichum 17 (GEO) – A secondary culture used to develop yeast or mould like growth on the surface of cheese.

Lipase – An enzyme used in cheese making to develop a piquant flavor.

MA4001/4002 – A common mesophilic culture in cheese making.

Mesophilic culture – Cheese cultures designed to be used in cheese making where the milk temperature does not exceed 40°C (104°F).

Penicillium candidum – A mould used to develop the typical white mould rind on brie or camembert style cheese.

Rennet – Traditionally made from an enzyme in a young mammal's stomach, rennet assists in coagulating the milk which allows the cheesemaker to separate curds and whey. Also available in vegetarian and microbial varieties.

Vegetable ash – See Charcoal Ash

Vegetarian rennet – A vegetarian version of animal rennet, used to assist in coagulating the milk to separate curds and whey.

White mould (PC) – See Penicillium Candidum

INDEX

Ashed Goat Logs 83

Belgian Mussels with Blue Cheese 127
Buffalo Brie 86
Burrata 98

Double-Baked Cheese Soufflé 132

Filo Cheese Triangles 152
French-style Brie 55
French-style Goat Cheese Salad 130
French Goat Crottin 69
French Onion Soup 137
French Sourdough with
    Caramelized Onions 118
Fresh Goat Curd 78
Fresh Herbed Goat Cheese 75
Fresh Herb Ricotta 44

German Pumpernickel Bread 105
Goat Cheese Crostada 138
Goat Cheese Mustard Gougères 135
Greek-Style Feta 58

Haloumi 50
Haloumi Potato Pie 140
Herb Focaccia 1115
Homemade Buffalo Mozzarella 93

Irish Fruit Bread 108
Italian Grissini 123

Lactic Blue Buffalo Cheese 89
Lactic Buffalo Cheese 88

Middle Eastern Pita Bread 120

Persian (marinated) Feta 76
Provençal Flatbread 107
Provençal Tomato & Goat Cheese Tarts 155
Pure Goat Brie 71

Raisin Rye Sticks with Bran & Walnuts 116
Ricotta-stuffed Eggplant Rolls 141
Risotto of Four Cheeses 143
Risotto with Spinach & Blue Cheese 146
Roasted Beetroot Salad with Persian Feta ... 145
Rosemary & Sultana Panini 113

Saganaki 129
Semi-Hard Goat Cheese with Peppercorns ... 81
Soft Blue Goat Cheese 65
Spanish-style Sheep's Cheese 95
Spreadable Cream Cheese 41
Stilton-Style Blue Cheese 61
Sweet Potato & Peanut Salad with Haloumi ....
    148

Tomato & Mozzarella Frittata 151
Traditional Morbier (firm washed rind) 52

Washed Rind Cheese (soft) 47
Whey Ricotta 45
Whole Milk Ricotta 43
Witlof Salad with Apples, Blue Cheese
    & Pecans 153

Zataar-Spiced Breads 110

First published in 2017 by New Holland Publishers
London • Sydney • Auckland

The Chandlery, 50 Westminster Bridge Road, London SE1 7QY, United Kingdom
1/66 Gibbes Street, Chatswood, NSW, 2067, Australia
5/39 Woodside Ave, Northcote, Auckland, 0627, New Zealand

newhollandpublishers.com

Copyright © 2017 New Holland Publishers
Copyright © 2017 in text: Tamara Newing
Copyright © 2017 in images: New Holland Publishers; Shutterstock page 121

All rights reserved. No part of this publication may be reproduced, stored in a retrieval system or transmitted, in any form or by any means, electronic, mechanical, photocopying, recording or otherwise, without the prior written permission of the publishers and copyright holders.

A record of this book is held at the British Library and the National Library of Australia.

ISBN 9781742579641

Group Managing Director: Fiona Schultz
Publisher: Monique Butterworth
Production Director: James Mills-Hicks
Project Editor: Gordana Trifunovic
Designer: Lorena Susak
Proofreader: Kaitlyn Smith
Photographer: Rochelle Seator
Stylist: Kirsty Bryson
Printer: Toppan Leefung Printing Ltd

10 9 8 7 6 5 4 3 2 1

Keep up with New Holland Publishers on Facebook
www.facebook.com/NewHollandPublishers